EUGENE RIMMEL

THE BOOK OF PERFUMES

Elibron Classics
www.elibron.com

Elibron Classics series.

© 2005 Adamant Media Corporation.

ISBN 1-4021-7820-4 (paperback)
ISBN 1-4021-3143-7 (hardcover)

This Elibron Classics Replica Edition is an unabridged facsimile
of the edition published in 1865 by Chapman & Hall,
London.

THE BOOK OF PERFUMES.

THE SHOP OF RENÉ THE PERFUMER, ON PONT
AU CHANGE, PARIS.

BOOK OF PERFUMES.

BY

EUGENE RIMMEL,

MEMBER OF THE SOCIETY OF ARTS, AND REPORTER OF THE JURY AT THE GREAT
EXHIBITION (PERFUMERY CLASS).

WITH

ABOVE 250 ILLUSTRATIONS BY BOURDELIN, THOMAS, ETC.

SECOND EDITION.

LONDON:

CHAPMAN AND HALL, 193, PICCADILLY.

TO BE HAD ALSO OF THE AUTHOR,

96, STRAND, 128, REGENT STREET, 24, CORNHILL, LONDON.
17, BOULEVARD DES ITALIENS, PARIS.

MDCCCLXV.

STEPHEN AUSTIN,

PRINTER, HERTFORD.

PREFACE.

LTHOUGH I am aware that Prefaces are out of fashion, and that it is now customary to plunge at once *in medias res*, I feel bound in this instance to deviate from the general rule, and to explain how I was led to emerge from the depths of my laboratory, and to appear thus before the public *in an entirely new character*. Four years ago, I had to prepare for the Society of Arts a paper on "The Art of Perfumery, its History and Commercial Development," and, to qualify myself for the task, I was forced to devour a huge pile of *big books* in order to ascertain through what mysterious arts the Ancients ministered to the gratification of their olfactory sense, and to the embellishment of "the human face divine." Two years later, I was called upon to form part of the Jury at the Great Exhibition, and to draw up the official report of the Perfumery class. The researches I had to make on the former occasion, and the observations I gathered on the latter, gave me a complete insight into the world of "sweet smells," both ancient and modern; and, thinking that the notes I had thus collected, combined with the

b

results of my experience as a practical perfumer, and of my rambles in foreign lands, might prove interesting to some readers, and especially to ladies, I published in the "Englishwoman's Magazine" a series of articles on the "History of Perfumery and the Toilet." These few stray leaves having met with a much more favourable reception than I had presumed to anticipate, I have been induced to re-publish them in the shape of a book, adding thereto a great deal of fresh matter, and numerous illustrations.

Many writers have already exercised their pen on the subject of Perfumery, from Aspasia, the wife of Pericles, to Mr. Charles Lilly, the Perfumer, of the Strand, at the corner of Beaufort Buildings, whose premises I have now the honour of occupying, and whose name was immortalized in the "Tatler," and other magazines of the period. The list of these works would be long and tedious, and those that are worth noticing will be found chronicled in their proper place in the following pages.

Modern books on Perfumery may be divided into two classes, some being simply books of recipes, laying claim to a useful purpose which, however, they do not fulfil, since they contain nothing but antiquated formulæ long discarded by intelligent practitioners; and others being what our neighbours call *réclames*, namely, works written in a high-flown style, but invariably terminating *en queue de poisson*, with the praise of some preparation manufactured by the author.

Besides these productions, articles on Perfumery

have occasionally appeared in periodicals; but though some of them are handled with evident talent, the want of technical knowledge on the part of the writers considerably impairs their value. I may mention as an instance an essay of that kind lately published by the "Grand Journal," in which a certain Parisian doctor gravely asserts that rouge is made of vermilion, and commits numerous other blunders, which may pass unnoticed by the general public, but which in the eye of a practical man, denote his utter ignorance of the subject he attempted to treat.

In writing this book, I have endeavoured to stéer clear of these objections, by adopting the following plan, which will be found to differ entirely from those of my predecessors. After devoting a few pages to the physiology of odours in general, I trace the history of perfumes and cosmetics from the earliest times to the present period, and that is the principal feature of my work. I then briefly describe the various modes in use for extracting the aromas from plants and flowers, and conclude with a summary of the principal fragrant materials used in our manufacture; in fine, I give all the information which I think likely to interest the general reader. The only recipes which I quote are those which I think, from their quaintness, likely to amuse, but I abstain from giving modern formulæ, for the following reasons, which I hope may appear sufficient:

There was a time when ladies had a private still-room of their own, and personally superintended the

various "confections" used for their toilet; but it
was then almost a matter of necessity, since native
perfumers were scarce, and exotic preparations expensive and difficult to procure. Such is not the
case now: good perfumers and good perfumes are
abundant enough; and, with the best recipes in the
world, ladies would be unable to equal the productions
of our laboratories, for how could they procure the
various materials which we receive from all parts of
the world? And were they even to succeed in so doing,
there would still be wanting the necessary utensils
and the *modus faciendi*, which is not easily acquired.
I understand the use of a cookery-book, for the
culinary art is one that *must* be practised *at home*;
but perfumery can always be bought much better
and cheaper from dealers, than it could be manufactured privately by untutored persons.

The recipes, therefore, admitting them to be genuine,
would only be of use to those who follow the same
pursuit as myself. But is it to be reasonably expected that, after spending my life in perfecting my
art, I am thus to throw away the result of my
labours in a fit of Quixotic generosity? Had I discovered some means of alleviating the sufferings of
my fellow-creatures, I should think myself in duty
bound to divulge my secret for the benefit of hu-
manity at large; but I do not feel impelled by the
same considerations to give to my rivals in trade the
benefit of my practical experience, for then, indeed,
"Othello's occupation" would be "gone." This may be

thought by some a selfish way of reasoning; but on due reflection, they will find that I am only more *sincere* than those who appear to act differently. As a proof, I need but point out the inconsistency of a perfumer who claims some superiority for his art in compounding, and who, at the same time, explains by what means he attains that superiority. Is he not at once destroying his *prestige* if he professes to enable others to manufacture just as well as himself? The conclusion of all this is, that the recipes given in books are never those actually used; and I say, therefore, *cui bono ?*

If I have avoided recipes, I have also shunned any allusions to my personal trade. As a man of business, I do not underrate the value of advertisements; but I like everything in its place, and consider this hybrid mixture of literature and puff, an insult to the good sense of the reader.

Before I close this brief address, I wish to acknowledge, with best thanks, the assistance I have received in the shape of very interesting notes, from many of my friends and correspondents, among whom I may mention, Mr. Edward Greey, of the Royal West India Mail Company; Mr. Chapelié, of Tunis; Mr. Thunot, of Tahiti; Mr. Schmidt, of Shang-Hae; Mr. Elzingre, of Manilla; Professor Müller, of Melbourne; Mr. Hannaford, of Madras; and last, not least, Mr. S. Henry Berthoud, the eminent French *littérateur*, who very kindly placed his unique museum at my disposal.

I have also found some valuable information in the

following books (besides others mentioned in the course
of the work) :—Sir Gardner Wilkinson's "Ancient
Egyptians;" Mr. Layard's "Nineveh;" Mr. Eastwick's
excellent translations of Sâdî's "Gulistān," and the
"Anvar-í Suhaili;" Mr. Monier Williams's no less
admirable adaptation of "Śakoontalá;" Consul Pethe-
rick's "Egypt, the Soudan, and Central Africa;" Dr.
Livingstone's "Travels," and Mr. Wright's "Domestic
Manners and Sentiments During the Middle Ages."
I have not the honour of knowing these authors, but
I hope they will excuse me for having borrowed
from them what belonged to my subject.

In conclusion, I crave for this offspring of my leisure
hours (which are but few), the same indulgence which
has been shown to the objects contained in the Work-
men's Exhibitions lately held in various parts of the
Metropolis, in which the labour and difficulty in producing an article is more taken into account than the
actual merit of the production. Mine is a plain, un-
varnished tale, without any literary pretension whatever; and if I have picked up a few gems on my way,
and inserted them in my mosaic work, I claim but to
be the humble cement which holds them together.

EUGENE RIMMEL.

96, STRAND, 15th December, 1864.

CONTENTS.

CHAPTER IV.

THE ANCIENT ASIATIC NATIONS.

CHAPTER V.

THE GREEKS.

CHAPTER VI.

THE ROMANS.

CHAPTER VII.

THE ORIENTALS.

CHAPTER VIII.

THE FAR EAST.

CHAPTER IX.

UNCIVILIZED NATIONS.

CHAPTER X.

FROM ANCIENT TO MODERN TIMES.

CHAPTER XI.

THE COMMERCIAL USES OF FLOWERS AND PLANTS.

CHAPTER XII.

MATERIALS USED IN PERFUMERY.

LIST OF ILLUSTRATIONS.

THE FLORAL WORLD.

THE BOOK OF PERFUMES.

CHAPTER I.

PHYSIOLOGY OF PERFUMES.

Ah, what can language do ? ah, where find words
Ting'd with so many colours; and whose powers,
To life approaching, may perfume my lays
With that fine oil, those aromatic gales,
That inexhaustible flow continual round ?—THOMSON.

MONG the many enjoyments ments provided for us by bountiful Nature, there are few more delicate and, at the same time, more keen than those derived from the sense of smell. When the olfactory nerves, wherein that sense resides, are struck with odoriferous emanations, the agreeable impression they receive is rapidly and vividly transmitted to the brain, and thus acquires somewhat of a mental character.

Who has not felt revived and cheered by the balmy
fragrance of the luxuriant garden or the flowery meadow ?   Who has not experienced the delightful sensa-
tions caused by inhaling a fresh breeze loaded with the
spoils of the flowery tribe?—that "sweet south," so
beautifully described by Shakspeare as

> "Breathing o'er a bank of violets,
> Stealing and giving odour."

An indescribable emotion then invades the whole being ;
the soul becomes melted in sweet rapture, and silently
offers up the homage of its gratitude to the Creator
for the blessings showered upon us; whilst the tongue
slowly murmurs with Thomson—

> " Soft roll your incense, herbs, and fruits, and flowers,
> In mingled clouds to Him whose sun exalts,
> Whose breath perfumes you, and whose pencil paints !"

It is when nature awakes from her long slumbers,
and shakes off the trammels of hoary Winter, at that
delightful season which the Italian poet so charmingly
hails as the "youth of the year,"

> "Primavera, gioventù dell' anno !"

that the richest perfumes fill the atmosphere.   The fair
and fragile children of Spring begin to open one by one
their bright corols, and to shed around their aromatic
treasures :—

> " Fair-handed Spring unbosoms every grace ;
> Throws out the snowdrop and the crocus first ;
> The daisy, primrose, violet darkly blue,
> And polyanthus of unnumber'd dyes ;
> The yellow wallflower, stained with iron-brown,
> And lavish stock that scents the garden round."

But soon—too soon, alas !—those joys are doomed to

pass ; like the maiden ripening into the matron, the flower becomes a seed, and its fragrance would for ever be lost, had it not been treasured up in its prime by some mysterious art which gives it fresh and lasting life.

> " The roses soon withered that hung o'er the wave,
> But some blossoms were gathered while freshly they shone,
> And a dew was distilled from their flowers that gave
> All the fragrance of summer when summer was gone."

Thus the sweet but evanescent aroma, which would otherwise be scattered to the winds of heaven, assumes a durable and tangible shape, and consoles us for the loss of flowers when Nature dons her mourning garb, and the icy blast howls round us. To minister to these wants of a refined mind—to revive the joys of ethereal spring by carefully saving its balmy treasures—constitutes the art of the perfumer.

When I say "the art of the perfumer," let me explain this phrase, which might otherwise appear ambitious. The first musician who tried to echo with a pierced reed the songs of the birds of the forest, the first painter who attempted to delineate on a polished surface the gorgeous scenes which he beheld around him, were both artists endeavouring to copy nature ; and so the perfumer, with a limited number of materials at his command, combines them like colours on a palette, and strives to imitate the fragrance of all flowers which are rebellious to his skill, and refuse to yield up their essence. Is he not, then, entitled to claim also the name of an artist, if he approaches even faintly the perfections of his charming models ?

The origin of perfumery, like that of all ancient arts, is shrouded in obscurity. Some assert that it was first discovered in Mesopotamia, the seat of earthly paradise, where, as Milton says,

> " Gentle gales,
> Fanning their odoriferous wings, dispense
> Native perfumes, and whisper whence they stole
> Those balmy spoils ;"

others that it originated in Arabia, which has long enjoyed, and still retains, the name of the "land of perfumes." Whatever may be the true version, it is evident that when man first discovered

> " What drops the myrrh, and what the balmy reed,"

his first idea was to offer up these fragrant treasures as

A Primitive Perfume Altar.

a holocaust to the Deity. The word perfume (*per*, through, *fumum*, smoke) indicates clearly that it was first obtained by burning aromatic gums and woods ; and it seems as if a mystic idea was connected with this mode of sacrifice, and as if men fondly believed that their prayers would sooner reach the realms of their gods by being wafted on the blue wreaths which slowly ascended to heaven and disappeared in the atmosphere, whilst their intoxicating fumes threw them into religious ecstasies. Thus we find perfumes form a part of all primitive forms of worship. The altars of Zoroaster and of Confucius, the temples of Memphis and those of Jerusalem, all smoked alike with incense and sweet scented woods.

Among the Greeks, perfumes were not only considered as a homage due to their deities, but as a sign of their presence. Homer and other poets of that period never mention the apparition of a goddess without speaking of the ambrosial clouds which surround her. Thus is Cupid's fair mother described in the "Iliad" when she visits Achilles :—

> " Celestial Venus hovered o'er his head,
> And roseate unguents heavenly fragrance shed ! "

And in one of Euripides' tragedies, Hippolites, dying, exclaims, " O Diana, sweet goddess, I know that thou art near me, for I have recognised thy balmy odour."

The use of perfumes by the ancients was not long confined to sacred rites. From the earliest times of the Egyptian empire we find that they were adapted to private uses, and gradually became an actual necessary to those who laid any claim to refined taste and habits. We may say that perfumery was studied and cherished by all the various nations which held in turn the sceptre of civilization. It was transmitted by the Egyptians to the Jews, then to the Assyrians, the Greeks, the Romans, the Arabs, and at last to the modern European nations, when they emerged from their long chaos of barbarous turmoil, and again welcomed the arts of peace. It will be our study to trace its course through these different phases; to dive into the mysteries of the toilet of the Greek beauty and the Roman matron; to describe the various ways in which ladies have endeavoured, at all times and in all countries, to increase and preserve the charms lavished upon them by nature; and, lastly, to record the progress of perfumery to the present

Egyptian Princess.

Powdered Belle of the last century.

period, when, having shaken off the trammels of ignorance and quackery, it aspires to become useful no less than ornamental. To render the history of the Toilet more complete, we shall bestow a passing glance on the sundry styles of dressing the hair at different periods,

from the Egyptian princess under the Cheops dynasty
to the powdered belle of the last century. Nor shall
civilised people monopolise our
whole attention : in our roamings
"all round the world," we shall
find even among barbarous tribes
some curious fashions to register,
and African beauties as well as
Tartar damsels will have to reveal
to us the secrets of their so-called
embellishments. We shall then
conclude with a brief description
of the principal modes used in
extracting perfumes from flowers

African Headdress.

and aromatic plants, of the chief materials to which
we are indebted for our aromatic treasures, and of the

Lepcha Headdress.

various substances which
might also be rendered
available for that purpose.

Before commencing,
however, this chronolo-
gical narration, I may be
allowed to say a few words
on odours in general.

All plants and all
flowers exhale an odour
more or less perceptible
—more or less agreeable.

Some flowers, like that of the orange-tree and the rose,
possess such a powerful aroma that it scents the air for

miles around. Those who have the good fortune to travel
in the "genial land of Provence," when the flowers are
in full bloom,

> "And the woodbine spices are wafted abroad
> And the musk of the roses blown,"

are saluted (as I have frequently been myself), with the

Floral Clock.

balmy breezes emanating from the floral plantations
of Grasse or Nice long before they reach them. Some
flowers have a stronger smell at sunrise, some at mid-
day, others at night. This depends in a great measure
on the time they are wont to open, which varies so much

among the fragrant tribe, that it has allowed a patient botanist to form a floral clock, each hour being indicated by the opening of a particular flower.

The accompanying illustration will give some idea of this floral clock. I have taken it from an old work on botany, but for its accuracy I cannot vouch. It consists of the following flowers, the hour stated for some being in the morning and for others in the evening :—

1	Rose.	5	Convolvulus.	9	Cactus.
2	Heliotrope.	6	Geranium.	10	Lilac.
3	Water-lily.	7	Mignonette.	11	Magnolia.
4	Hyacinth.	8	Carnation.	12	Violet and Pansy.

All odours are not alike in intensity. Some flowers lose their fragrance as soon as they are culled; others, on the contrary, preserve it even when dried. None, however, can equal in strength and durability the odours derived from the animal kingdom. A single grain of musk will retain its aroma for years, and impart it to everything with which it comes in contact.

Odours have been classified in various ways by learned men. Linnæus, the father of modern botanical science, divided

Linnæus, the Botanist.

them into seven classes, three of which only were pleasant odours, viz., the aromatic, the fragrant, and the ambrosial: but, however good his general divisions may have been, this classification was far from correct, for he placed carnation with laurel leaves, and saffron with jasmine, than which nothing can be more dissimilar. Fourcroy divided them into five series, and De Haller into three. All these were, however, more theoretical than practical, and none classified odours by their resemblance to each other. I have attempted to make a new classification, comprising only pleasant odours, by adopting the principle that, as there are primary colours from which all secondary shades are composed, there are also primary odours with perfect types, and that all other aromas are connected more or less with them.

The types I have adopted will be found in the following table:—

CLASSIFICATION OF ODOURS.

CLASSES.	TYPES.	ODOURS BELONGING TO THE SAME CLASS.
Rose	Rose	Geranium, Sweetbriar, Rhodium, Rosewood
Jasmine	Jasmine	Lily of the Valley
Orange Flower	Orange Flower	Acacia, Syringa, Orange leaves.
Tuberose	Tuberose	Lily, Jonquil, Narcissus, Hyacinth.
Violet	Violet	Cassie, Orris-root, Mignonette.
Balsamic	Vanilla...................	Balsam of Peru and Tolu, Benzoin. Styrax, Tonquin Beans, Heliotrope.
Spice....................	Cinnamon	Cassia, Nutmeg, Mace, Pimento,
Clove....................	Clove	Carnation, Clove Pink.
Camphor	Camphor	Rosemary, Patchouly.
Sandal	Sandalwood	Vetivert, Cedarwood.
Citrine	Lemon	Bergamot, Orange, Cedrat, Limette
Lavender...............	Lavender...............	Spike, Thyme, Serpolet, Marjoram.
Mint	Peppermint...........	Spearmint, Balm, Rue, Sage.
Aniseed	Aniseed	Badiane, Carraway, Dill, Coriander, Fennel.
Almond	Bitter Almonds	Laurel, Peach Kernels, Mirbane.
Musk.................. ...	Musk....................	Civet, Musk-seed, Musk-plant.
Amber	Ambergris	Oak-moss.
Fruit.....................	Pear	Apple, Pine-apple, Quince.

This is the smallest number of types to which I could reduce my classification, and even then there are some particular odours, such as that of winter-green, which it would be difficult to introduce into either class ; nor does this list comprise the scents which are produced by blending several classes together.

Jean Jacques Rousseau, Zimmermann, and other authors, say that the sense of smell is the sense of imagination. There is no doubt that, as I have observed before, pleasant perfumes exercise a cheering influence on the mind, and easily become associated with our remembrances. Sounds and scents share alike the property of refreshing the memory, and recalling vividly before us scenes of our past life—an effect which Thomas Moore beautifully illustrates in his " Lalla Rookh :"—

> "The young Arab, haunted by the smell
> Of her own mountain flowers as by a spell,
> The sweet Elcaya, and that courteous tree,
> Which bows to all who seek its canopy,
> Sees call'd up round her by those magic scents
> The well, the camels, and her father's tents;
> Sighs for the home she left with little pain,
> And wishes e'en its sorrows back again."

Tennyson expresses the same feeling in his " Dream of fair women."

> " The smell of violets, hidden in the green,
> Pour'd back into my empty soul and frame
> The times when I remember to have been
> Joyful and free from blame."

Criton, Hippocrates, and other ancient doctors, classed perfumes among medicines, and prescribed them for many diseases, especially those of a nervous kind.

Pliny also attributes therapeutic properties to various aromatic substances,[1] and some perfumes are still used in modern medicine.

Discarding, however, all curative pretensions for perfumes, I think it right, at the same time, to combat the doctrines of certain medical men who hold that they are

injurious to health. It can be proved, on the contrary, that their use in moderation is more beneficial than otherwise ; and in cases of epidemics they have been known to render important service, were it

[1] Pliny, in his Natural History, mentions eighty-four remedies derived from rue, forty-one from mint, twenty-five from pennyroyal, forty-one from the iris, thirty-two from the rose, twenty-one from the lily, seventeen from the violet, etc.   (Pliny's Nat. Hist. b. xx. and xxi.)

only to the four thieves who, by means of their famous aromatic vinegar,[1] were enabled to rob half the population of Marseilles at the time of the great plague.

It is true that flowers, if left in a sleeping-apartment all night, will sometimes cause headache and sickness, but this proceeds not from the diffusion of their aroma, but from the carbonic acid they evolve during the night. If a perfume extracted from these flowers were left open in the same circumstances, no evil effect would arise from it. All that can be said is that some delicate people may be affected by certain odours; but the same person to whom a musky scent would give a headache might derive much relief from a perfume with a citrine basis. Imagination has, besides, a great deal to do with the supposed noxious effects of perfumes. Dr. Cloquet, who may be deemed an authority on this subject, of which he made a special study, says in his able Treatise on Olfaction :—" We must not forget that there are many effeminate men and women to be found in the world who *imagine* that perfumes are injurious to them, but their example cannot be adduced as a proof of the bad effect of odours. Thus Dr. Thomas Capellini relates the story of a lady who *fancied* she could not bear the smell of a rose, and fainted on

---

[1] It is related that during the great plague which visited Marseilles four robbers, who had become associated, invented an aromatic vinegar by means of which they could rob the dead and the dying, without any fear of infection. This vinegar was long known in France under the name of " Vinaigre des quatre Voleurs," and gave the first idea of Toilet Vinegar.

receiving the visit of a friend who carried one, and yet the fatal flower was only *artificial.*" [1]

Were any other argument wanting to vindicate perfumes from the aspersions cast upon them, I would say that we are prompted by a natural instinct to seek and enjoy pleasant odours, and to avoid and reject unpleasant ones, and it is unreasonable and unjust to suppose that Providence has endowed us with this discerning power, to mislead us into a pleasure fraught with danger, or even discomfort.

[1] Osphrésiologie, ou Traité de l'Olfaction, par le Dr. H. Cloquet, chap. v. p. 80.

AN EGYPTIAN TEMPLE.

CHAPTER II.

THE EGYPTIANS.

The barge she sat in, like a burnished throne,
    Burnt on the water; the poop was beaten gold,
Purple the sails, and so perfumed that
    The winds were love sick . . . .

ANTONY AND CLEOPATRA.

ONG before any other nation, Egypt had learned, or rather invented, the art of raising lofty temples to its gods, magnificent palaces to its princes, and immense cities for its people, and of decorating them with all the various treasures which nature had placed at its disposal. Whilst the Jews and other surrounding people were confined to the simplicities of pastoral life, the Egyptians were enjoying the luxuries of refinement, and carried them to an extent which was not surpassed, if equalled, by those who, after them, successively held the sceptre of civilization.

Although the Egyptians left no trace of their literature, the ample descriptions given by the Greek and Latin authors, the frequent mention made in the Bible and, above all, the numerous paintings and sculptures found on their monuments and in their tombs, give us a complete insight into their manners and mode of life. The huge piles of granite which they reared over the last asylum of their monarchs, in the vain hope of securing their eternal peace, and of screening them from the profane gaze of intruders, were not proof against the cupidity of the *fellahs* of modern Egypt, who found their way into the abodes of the dead in search of the treasures buried with them. This unholy spoliation was not, however, entirely barren of happy results,

> " For nought so vile on the earth doth live
> But to the earth some special good doth give."

In this instance the inroads made by the avaricious

Mummy Pit.

plunderers into the ruined palaces and mummy pits paved the way for equally daring but more disinterested explorers, and enabled scientific men like Sonnini, Belzoni, Savary, Champollion, Sir Gardener Wilkinson, Mariette, and others, to dive into the mysteries of ancient Egyptian customs, and to give us a correct and vivid account of what the world was long before the era of written history. We learn from these descriptive illustrations, confirmed by the records of ancient writers and by the numerous implements found intact in the tombs,.

at perfumes were extensively consumed in Egypt, and applied to three distinct purposes—offerings to the gods, embalming the dead, and uses in private life.

At all the festivals held by the Egyptians in honour of their numerous deities, perfumes played a conspicuous part, and they also ranked among the most grateful of their daily oblations. With the *naïve* gratitude of a primitive people, they felt it a sort of duty to offer the finest fruit, the fairest flower, the richest wine, the fattest bullock, to the gods, who were supposed to have dispensed those boons ; but of all other sacrifices that of incense appeared to them the most refined and appropriate. In the temples of Isis, the good "goddess ;" of Osiris, the eternal rival of Typhon ; of Pasht, or the Egyptian Diana ; aromatic gums and woods were constantly burned by the priests, and on grand state occasions the king himself officiated,

Rhamses III. Sacrificing.

holding a censer in one hand, and in the other a small

2

vase with a spout containing wine or perfumed oil for
libations to be poured on the altar. The engraving on
the preceding page, which represents Rhamses III.,
illustrates this mode of sacrifice.

In ordinary ceremonies incense alone was offered, in
the shape of round balls or pastilles, which were thrown
into the censers. Those censers were not swung about,
as are those used in Catholic churches: they were
straight, and held firmly in the right hand, whilst the
incense was thrown in with the left, an operation which
must have required some little practice, if performed as
adroitly as the Egyptian painters would lead us to
believe.

Egyptian Censers.

At Heliopolis, the City of the Sun, where the great
orb was adored under the name of Re, they burned
incense to him three times a day—resin at his first
rising, myrrh when in the meridian, and a mixture of
sixteen ingredients, called *Kuphi*, at his setting.

The sacred bull, Apis, had also his share of such
homage. Those who wished to consult him burnt incense on his altar, filled the lamps which were lighted
there with fragrant oils, and deposited a piece of money
before the statue of the god. They then whispered
softly to him the question they wished to ask, and
issued from the temple carefully stopping their ears.

The first word that was uttered by any one they chanced to meet after that, was taken by them to convey the reply which they sought.

Besides incense, ointment was also offered to the gods, and formed an indispensable part of what was considered a complete oblation. It was placed before the deity in vases of alabaster or other costly material, on which was frequently engraved the name of the god to whom it was offered. Some-

Offerings of Ointment.

times the king or the priest took out a certain portion, and anointed the statue of the divinity with his little finger.

At the *fête* of Isis, which was performed with great magnificence, they sacrificed an ox filled with myrrh, frankincense, and other aromatic substances, which they burnt, pouring a quantity of oil over it during the process. The fragrant vapours thus produced counteracted the smell of the burning flesh, which would otherwise have been unbearable, even to the most ardent votaries of the goddess.

The two principal festivals in honour of Osiris were held at six months' distance from each other. The first was meant to commemorate the loss, and the second the finding, of Egypt's tutelar god. At the latter the priests carried the sacred chest, inclosing a small vessel of gold, into which they poured some water, and all the people assembled cried out, "Osiris is found!"

They then threw into the water some fresh mould, together with rich odours and spices, and shaped it into a little image resembling a crescent, which was supposed to typify the essence and power of earth and water.

It was, however, in their grand religious processions that they made the most luxurious display of perfumes. In one of those described as having taken place under one of the Ptolemies, marched one hundred and twenty children, bearing incense, myrrh, and saffron in golden basins, followed by a number of camels, some carrying three hundred pounds weight of frankincense, and others a similar quantity of crocus, cassia, cinnamon, orris, and other precious aromatics.

No king could be crowned without being anointed: this was done privately by the priests, who pretended that the ceremony had been performed by a god, in order to convey to the people a more exalted notion of the benefits conferred on their monarchs. The latter also shared with the deities the privilege of being offered incense: but this only on special occasions, such as their return from a victorious campaign. The king then entered the capital, borne in his chair of state, and accompanied by a brilliant *cortége*. A long procession of priests came to meet him, dressed in gorgeous robes, and holding censers full of incense, whilst a sacred scribe read from a papyrus roll the glorious deeds of the victorious sovereign.

The Egyptians believed in the transmigration of souls —a doctrine afterwards adopted by Pythagoras and other Grecian philosophers. They held that, after leaving the body of a man, his soul entered that of some other animal, and, having successively passed through all creatures of the earth, water, and air, it again assumed the human shape, which journey was accomplished in the lapse of three thousand years. This belief would account for the very great care they took in embalming the bodies of their dead, so that, after having concluded their long journey, the souls might find their original envelopes in a tolerable state of preservation. Diodorus, however, assigns another reason for this custom, and says the wealthy Egyptians kept the bodies of their ancestors in magnificent rooms set apart for that purpose, in order to have the gratification of contemplating the features of those who had died many generations before them, for the whole appearance of the person was so well preserved that it could be easily recognised.

Several times during the year these mummies were brought out and received the greatest honors. Incense and libations were offered to them, and sweetly scented oil was poured over their heads and carefully wiped off with a towel carried on the shoulder for the pur-

Priest pouring Oil over a Mummy.

pose. A priest was generally called in to officiate on these occasions.

The operation of embalming was performed in the following manner by the ancient Egyptians, according to Herodotus:—They first extracted the brains through the nostrils by means of a curved iron probe, and filled the head with drugs; then, making an incision in the side with a sharp Ethiopian stone, they drew out the intestines, and inserted into the cavity powdered myrrh,

Embalming Mummies (Perfuming the Body).

cassia, and other perfumes, frankincense excepted. After sewing up the body they kept it in natron[1] for seventy days, and then wrapped it up entirely with bands of fine linen, smeared with gum, and laid it in a wooden case, made in the ape of a man, which they laced upright against the all.

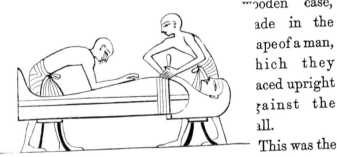

Embalming Mummies (Binding the Body).

This was the st class, or "Osiris style," of embalming; but, being very expensive, it was confined to the richest people. Another

[1] A native sesquicarbonate of soda found in great quantities in Egypt.

mode consisted in injecting oil of cedar into the body, without removing the intestines; whilst, in the case of the poorer class of people, the body was merely cleansed with syrmœa and salt, subjecting it, in both cases, to a natron bath, which completely dried the flesh. The first kind of embalming cost a talent, or about £250, the second twenty-two minæ, or £60, and the third was extremely cheap. These operations were

Embalming Mummies (Painting the Case).

performed by some persons regularly appointed for the purpose, and at Thebes there was a whole quarter of the town devoted to the preparation of the necessary implements. One of the most curious parts of the performance was that the *paraschistes*, or dissector, who had to make an incision in the body, ran away as soon as it was done, amid the bitter execrations of all those present, who pelted him unmercifully with stones, to testify their abhorrence of any one inflicting injury on a human creature, either alive or dead.

In some of the mummies the viscera were returned into the body, after being cleansed with palm wine and mixed with pounded aromatics; but for persons of distinction they placed the internal parts in four sepulchral vases, dedicated to different deities. The first jar, surmounted with a human head, was consecrated to Am-Set, a genius presiding over the South, and contained the large intestines; the second vase, covered with a *cynocephalus*, held the smaller viscera, and was dedicated to Ha-Pi, the genius of the North; the third, represented

Funeral Vase.

here, received the heart and lungs, and was decorated with a jackal's head, in honour of Traut-mutf, the genius of the East; and in the fourth, ornamented with a hawk's head, were deposited the liver and gall-bladder, under the protection of Krebsnif, the genius of the West, who was, as well as the three others, a son of Osiris. All these vases were filled with perfumes, to insure the preservation of their contents.

Embalming was not confined to the human species. Some animals, and principally those held sacred by the Egyptians, equally shared this privilege. When the divine bull, Apis, had completed the twenty-five years which were allotted to him as the extent of his natural life, the priests drowned him in the Nile, embalmed him, and buried him with great solemnity. Cats and other animals were also embalmed, and there are numerous specimens of their mummies in the British Museum.

In some barren parts of Egypt, where sand was more plentiful than aromatics, they preserved their dead by exposing them for some time on the ground to the burning rays of the sun, which completely desiccated the body. Sonnini describes, in his travels, a somewhat similar process carried on at a Capuchins' convent in the neighbourhood of Palermo, by means of which the bodies of all the community have been kept since its foundation by *broiling them over a slow fire*, forming, as he says, a most ghastly collection.

Mummy of a Cat.

Among many customs derived by modern Egyptians from their ancestors is that of embalming, which is still observed among wealthy people, and which, according to Maillet, is performed in the following manner : they wash the body several times with rose-water, perfume it with incense, aloes, and a variety of spices, wrap it up in a sheet moistened with liquid odours, and bury it with the richest suit of clothes belonging to the deceased.

Great as was the consumption of perfumes in Egypt for religious rites and funeral honours, it was scarcely equal to the quantity of aromatics used for toilet purposes. The Egyptians were very cleanly in their habits, and were the inventors of that complete system of baths which the Greeks and Romans borrowed from them, and which has remained in use among modern

Eastern nations. After the copious ablutions in which they indulged, they rubbed themselves all over with fragrant oils and ointments. This practice may appear repulsive to English readers, but it was, no doubt, required by the climate to give elasticity to the skin and counteract the effects of the sun. It is still generally kept up in Africa and other hot countries. The unguents used were of great variety, and were at first dispensed by the priests, who were then alone acquainted with the mysteries of the compounding art, and may be termed the first *manufacturing perfumers.* Some were flavoured with origanum, bitter almond, or other aromatics indigenous to the Egyptian soil; but the greater part of their ingredients, such as myrrh, frankincense, etc., came from Arabia. They

were kept in bottles, vases, or pots, made of alabaster, onyx, glass, porphyry, or other hard substances; and also in boxes made of carved wood or ivory, which assumed sometimes the most curious shapes, such as that of fishes, birds, etc. Some of these boxes were divided into compartments, like the specimen represented on the next page, which

Alabaster Vase containing Ointment.

probably held different cosmetics for the toilet. The preparation of those ointments was so perfect that a specimen in the Alnwick Castle museum has retained its scent after a lapse of three or four thousand years. They were generally very expensive, and the poorer classes, who could not afford such luxuries, used,

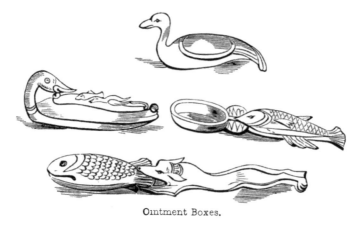

Ointment Boxes.

as a substitute, castor-oil, which Egypt produced in abundance.

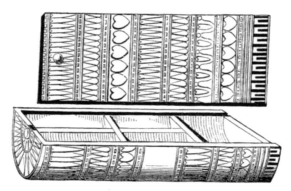

Ointment Box with compartments.

The perfumes and cosmetics resorted to by an Egyptian beauty of that period to heighten the effect of her charms were as numerous, if not as elegant, as those used by a modern votary of fashion, if we may

judge by the annexed toilet-case, containing a goodly
array of jars and bottles, supposed
to have belonged to a Theban lady.
Besides scented oils and unguents,
they used red and white paint for
their faces, and a black powder
called *kohl*, or *kohol*, made of antimony, which, applied with a wooden
or ivory bodkin to the pupils of the

A Theban Lady's Dressing-
case.

eyes, increased their brilliancy and made them appear

Kohl Bottles and Bodkin.

larger—a custom still prevalent throughout the East.
This kohol was held in vases of a curious
shape of which quantities have been found
in the tombs. One of those represented here
is evidently of Chinese origin, which leads
some people to suppose that the intercourse
between Egypt and the Celestial Empire
commenced at a very early date. This is,
however, a vexed question, on which many
large folios have been written, and I shall,
therefore, abstain, with wholesome dread, from offering

Chinese
Kohol bottle.

an opinion on such a controverted subject. I must not omit from the list of artifices employed by Egyptian belles, that of staining their fingers and the palms of their hands with the leaves of the henna (*Lawsonia iner-mis*), a practice which is supposed by some to have given rise to the Greek metaphor of "rosy-fingered Aurora."

The accompanying outline, taken from a painting at Thebes, represents an Egyptian lady at her toilet, and

An Egyptian Lady at her Toilet.

may convey an idea of the manner in which this important duty was performed. One of her attendants is pouring water over her, another rubs her with her hand, a third gives her to inhale the flower of the lotus, whilst the fourth is preparing to replace her ornaments.

Among the numerous toilet implements found in Egyptian tombs, the most conspicuous are mirrors and combs. The former were made of copper mixed with other metals, and their workmanship and polish were so excellent that some of them which have been revived after lying buried for many centuries equal almost in lustre our modern looking-glasses.   They were always

Egyptian Mirrors.

of a round shape, and the handle represented various
subjects, such as a deity, a flower, or sometimes a

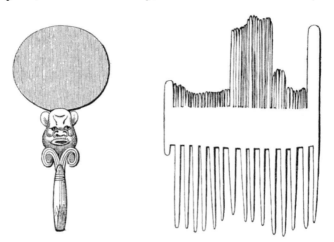

Egyptian Mirror with Typhonian handle.   Egyptian Comb.

typhonian monster, whose ugliness was calculated to set
off agreeably the lovely features reflected above it.

Egyptian combs were generally made of wood; some plain, and others carved. The annexed specimen is not unlike our modern small-tooth comb in shape.

The taste for perfumes and cosmetics went on increasing in Egypt until the time of Cleopatra, when it may be said to have reached its climax. This luxurious queen made a lavish use of aromatics, and it was one of

Cleopatra on the Cydnus.

the means of seduction she brought into play at her first interview with Mark Antony on the banks of the Cydnus, which is so beautifully described by Shakspeare. Glowing as the picture may seem, it is in no way overdrawn, and has been copied by our great poet,

almost word for word, from Plutarch's original recital,
to which he only added the charm of his verse.

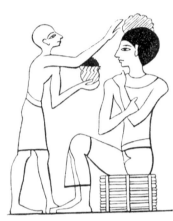

Slave Anointing Guest.

At all private festivals
perfumes were in great request. The first duty of
the attendant slaves on
the arrival of the visitors
was to anoint their heads,
or, rather, their wigs, for
they were all shaven and
wore this artificial covering, which served the purpose of modern turbans—
to protect them against
the rays of a scorching sun. During the entertainment, fresh flowers were used in great profusion;
chaplets of lotus decorated the necks of the guests,
garlands of crocus and saffron encircled the wine-cup,
floral wreaths were hung all round the room, and
over and under the tables were strewn various flowers
mingling their fragrance with the fumes of numerous
cassolettes, whilst, to leave no sense ungratified, musicians charmed the ear with the sweetest melodies. It
was thus that Agesilaus was received when he visited
Egypt; but the rude Spartan, unaccustomed to such
luxuries, refused the sweetmeats, confections, and perfumes, for which act of barbarism the polished natives held him in great contempt, as a man incapable
and unworthy of enjoying the refinements of good
society.

Herodotus relates a very curious custom which was observed at these Egyptian festivals. When the revel was at its height, a man entered, bearing the wooden image of a dead body, perfectly carved and painted, and cried aloud, "Look at this, drink and make merry, for so you will be after your death." Our modern "sensation"

An Egyptian Banquet.

dramatists could not wish for a better contrast, and I do not see, after all, why this strange habit should be more wondered at than the *fureur* with which they have sought lately to introduce *ghosts* into our public and private entertainments.

3

The Egyptians, as I said before, shaved their heads and chins, and looked with abhorrence on the rough-

Egyptian Barbers.

haired and long-bearded Asiatic nations. They only allowed their hair and beard to grow when in mourn-

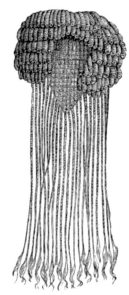

ing, and looked upon it in any other circumstance as a sign of low and slovenly habits. Most of them wore over their shaven polls wigs made of curled hair, with a series of plaits at the back, like the annexed specimens, one of which is taken from the British Museum, and the other from the Berlin collection of antiquities. Poor people, who could not afford the expense of real hair, had theirs made of black sheep's wool. By a singular contradiction, the great people wore artificial beards. which they likewise affixed to the

Egyptian Wig,
from the Berlin collection.

images of their gods. The beard of an individual of

Egyptian Wig in the British Museum (back and front view).

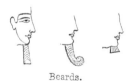

Beards.

rank was short and square, that of a king equally square, but much longer, and that of a god was pointed and turned up at the end.

Ladies wore their hair long, and worked into a multitude

The Lotus style]   Egyptian Ladies' Headdresses.   [The Peacock style.

of small plaits, part of which hung down their back and

the remainder descended on each side of the face, covering the ears completely. They generally had an ornamental

Egyptian Headdress from a Mummy case.

fillet round the head, with a lotus bud in front by way of a *ferronière*. Some of the *crême de la crême* indulged in a head-dress representing a peacock, whose gorgeous plumage set off their dark tresses; and princesses were usually dis-tinguished by a coiffure of extraordinary dimensions, com-bining all the riches of the animal, vegetable, and mineral kingdoms.

Modern Egypt has preserved many of the customs of its former inhabitants, on which I shall further descant when treating of the "Orientals." At present I shall proceed in due chronological order, and devote my next chapter to the Jews.

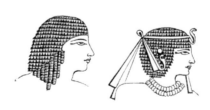

A GARDEN IN THE HOLY LAND.

CHAPTER III.

THE JEWS.

"Ointment and perfume rejoice the heart." — PROVERBS xxvii. 9.

LTHOUGH the Jews are undoubtedly the most ancient people extant, and the Holy Scriptures furnish us with abundant details respecting them since the commencement of the world, I have given them the second place in my history of Perfumes, because those luxuries do not appear to have come into general use among them until their return from Egypt. During their long captivity in that highly civilised country, they became initiated in all the refinements of their masters, being gradually transformed from a simple, pastoral people to a polished, industrious nation; and among the many arts which they brought back with them into their own country was that of perfumery.

Long before that time, however, they had probably discovered the aromatic properties of some of their native gums, and, prompted by that natural instinct to which I have already alluded, they had offered those fragrant treasures on the altars raised to their God. Thus we find Noah, on issuing from the ark, expressing his gratitude to the Almighty for his wonderful preservation by a sacrifice of burnt offerings, composed of "every clean beast and every clean fowl."[1] It is true that Genesis does not mention incense as having formed part of the holocaust, but the very words that follow, "And the Lord smelled a sweet savour," may lead us to assume that such was the case.

The mountains of Gilead, a ridge running from Mount Lebanon southward, on the east of the Holy Land, were covered with fragrant shrubs. The most plentiful among them was the amyris, which yields a gum known under the name of "balm of Gilead." Strabo also speaks of a field near Jericho, in Palestine, which was full of these balsam-trees. This gum seems to have formed an article of commerce at a very early period, for the Ishmaelite merchants to whom Joseph was sold by his brethren "came from Gilead with their camels, bearing spicery, and balm, and myrrh, going to carry it down to Egypt."[2]

Among the many commands which Moses received

[1] Genesis viii. 20. [2] Genesis xxxvii. 25.

from the Lord on his return from the land of captivity, were those of erecting the altar of incense, and compounding the holy oil and perfume:—

"And thou shalt make an altar to burn incense upon: of shittim wood shalt thou make it."

"And thou shalt overlay it with pure gold, the top

The Altar of Incense.

thereof, and the sides thereof round about, and the horns thereof; and thou shalt make unto it a crown of gold round about." [1]

In the same chapter we find the directions for making the holy anointing oil:—

"Take thou also unto thee principal spices, of pure

[1] Exodus xxx. 1–3.

myrrh five hundred shekels, and of sweet cinnamon half so much, even two hundred and fifty shekels, and of sweet calamus two hundred and fifty shekels. And of cassia five hundred shekels, after the shekel of the sanctuary, and of olive oil an hin.

" And thou shalt make it an oil of holy ointment, an ointment compound after the art of the apothecary (or *perfumer*) : it shall be a holy anointing oil." [1]

This oil served to anoint the tabernacle, the ark of the testimony, the altar of burnt offerings, the altar of incense, the candlesticks, and all the sacred vessels. It was also used to consecrate Aaron and his sons, conferring upon them perpetual priesthood from generation to generation. The ceremony was confined to the high priest, and was performed by pouring oil on the head in sufficient quantity to run down on the beard and the skirts of the garments. [2] There is a controversy as to when this practice was discontinued—some of the rabbis pretending that it was given up about fifty years before the destruction of the temple; while Eusebius is of opinion that it remained in use until our Saviour's time. [3]

Jewish kings were also anointed, but opinions differ very much as to whether it was done with the holy oil or common oil. Talmudic writers maintain that it was the peculiar privilege of the kings of the family of David to be anointed with the same holy oil which was used in the consecration of the high priest; but this can scarcely agree with the directions contained in

[1] Exodus xxx. 23, 24, 25.　　　[2] Psalm cxxxiii. 2.
[3] Eusebius, Demonstr. Evang. viii.

Exodus, by which the use of the holy ointment is confined to Aaron and his generation, to the exclusion of every other person.[1]

Although the ingredients of this oil are given to us, we are not told how it was prepared; and it seems difficult to understand how so many solid substances could be introduced into an hin of oil (which, according to Bishop Cumberland, is only a little more than a gallon) without destroying its liquidity. Maimonides pretends to explain this by saying that the four spices were pounded separately, then mixed together, and a strong decoction of them made with water, which, being strained from the ingredients, was boiled up with the oil till all the water had evaporated.[2]

The instructions given to Moses for compounding the holy incense were as follow:—

"Take unto thee sweet spices, stacte, and onycha, and galbanum; these sweet spices with pure frankincense; of each shall there be a like weight: and thou shalt make it a perfume, a confection after the art of the apothecary (or *perfumer*), tempered together pure and holy."[3]

The word *perfumer* occurs in some of the translations instead of that of *apothecary*, which is easily accounted for by the fact that in those times both callings were combined in one.

There is a great difference of opinion among scriptural commentators as to the true nature of stacte, onycha, and galbanum.

[1] Exodus xl. 13–15,   [2] De Apparatu Templi, cap. i. sec. 1.
[3] Exodus xxx. 34, 35.

Stacte, in Hebrew נָטָף (nataph), means dropping; and the Greek translation, Στακτὴ (staktè), has the same signification, hence it was thought by some to be storax and by others opobalsamum. Gesenius simply calls it a fragrant gum; but Professor Lee maintains it was myrrh, and he is probably correct. Rosenmüller, however, says that στακτὴ is derived from στάζειν, to distil, and that it was a distillate from myrrh and cinnamon. The word stacte also occurs in Latin authors, but their definitions do not agree; Pliny saying it is the natural exudation of the myrrh tree before it is cut, whilst Dioscorides pretends it is an unguent made of myrrh pounded in a little water and mixed with origanum.

There is still a greater controversy respecting onycha. Geddes and Boothroyd assimilate it to bdellium,[1] and Bochartus brings forth many arguments to prove it to have been labdanum,[2] one of the principal aromatics used by the Arabians. Maimonides states it was the hoof or claw of an animal, and Jarchi the root of a plant. The most general version, however, is that it was the shell of a fish found in the marshes of India, and that it derived its fragrance from the spikenard, upon which it fed. This fish was also found in the Red Sea, whence the Jews probably obtained it; and its white and transparent shell resembled a man's nail, which accounts for its name.[3]

Galbanum, in Hebrew חֶלְבְּנָה (chalbaneh), means unctuous, and was evidently a balsam. Bishop Patrick

---

[1] Gum-resin produced by the *balsamodendron mukul.*
[2] Gum of the *cistus creticus.*
[3] ὄνυξ (*onyx*) in Greek means a human nail.

says it must not be confounded with the common galbanum used in medicine, which has anything but an agreeable smell, but that it was a superior sort found in Syria, on Mount Amonus.

The word *tempered* has also been discussed, some pretending that it meant *salted*. Maimonides says that the incense was always mixed with salt of Sodom; but Bishop Horsley thinks that *tempered* in this case signifies *dissolved*.

Bezaleel and Aholiab, who were expert "in all manner of workmanship," were intrusted with the task of preparing the holy oil and incense, and it was strictly forbidden to use them for any other but sacred purposes.

" Whosoever shall make like unto that, to smell thereto, shall even be cut off from his people." [1]

The High Priest offering Incense.

It was likewise the exclusive prerogative of priests to

[1] Exodus xxx. 38.

offer up incense in the temple; and for having violated
this law, and disregarded the threats of Moses and
Aaron, Korah, Dathan, and Abiram, with two hundred
and fifty princes of the assembly, were swallowed up
by the earth, with their families and their goods.[1]

At a later period, King Uzziah was likewise repri-
manded by Azariah and eighty other priests for attempting to burn incense in the temple; and having persisted
in his design, he was struck with leprosy on the spot.[2]

The very severe penalties decreed by Moses against
any persons attempting to use the holy oil and incense
for private purposes, or even to compound similar preparations, give a very evident proof that the Jews had
brought from Egypt with them the habit of employing
perfumes, otherwise such prohibitions would have been
unnecessary.

With these they had also imported the cleanly habits
of the Egyptians, and that complete system of baths
which gave, as it were, new life to the frame, and which
naturally led them to the use of sweet unctions.

The purifications of women, as ordained by law, also
caused a great consumption of aromatics. They lasted
a whole year, the first six months being accomplished
with oil of myrrh, and the rest with other sweet odours.
This was the ordeal Esther had to undergo before she was
presented to king Ahasuerus, and "she obtained grace
and favour in his sight more than all the virgins."[3]

Perfumes were also one of the means of seduction
resorted to by Judith when she went forth to seek

[1] Numbers xvi. 32–35.   [2] 2 Chron. xxvi. 16–19.   [3] Esther ii. 12, 17.

Holofernes in his tent, and liberate her people from his oppression.

"She pulled off the sackcloth which she had on, and put off the garments of her widowhood, and washed her body all over with water, and anointed herself with precious ointment, and braided the hair of her head and put a tire upon it, and put on her garment of

Judith Preparing to meet Holofernes.

gladness, wherewith she was clad during the life of Manasses her husband."

"And she took sandals upon her feet, and put about her bracelets, and her chains and her rings, and her ear-rings and all her ornaments, and decked herself bravely, to allure the eyes of all men that should see her."[1]

[1] Judith x. 3, 4.

Perfumes were then very costly, and the Jews held them in such high esteem that they formed part of the presents made to sovereigns, as we find it to have been the case when the queen of Sheba visited king Solomon, and brought him "such spices as had never been seen." We also read that Hezekiah, receiving the envoys of the king of Babylon, showed them all his treasures, "the gold and silver, and the spices and sweet ointment." [1]

The most complete description of the various aromatics used by the Jews is to be found in the Canticles. A symbolical meaning has been ascribed, it is true, to this splendid Hebrew poem; but, even if taken in a figurative sense, the frequent mention of perfumes made in it shows that they must have been well-known and appreciated at the Jewish court.

"Because of the savour of thy good ointment, thy name is as good ointment poured forth."

"While the king sitteth at his table, my spikenard sendeth forth the smell thereof."

"My beloved is unto me as a cluster of camphire in the vineyards of Engedi."

"Who is this that cometh out of the wilderness like pillars of smoke perfumed with myrrh and frankincense, with all powders of the merchant?"

"The smell of thy garments is like the smell of Lebanon."

"Thy plants are an orchard of pomegranates, with pleasant fruits; camphire, with spikenard, spikenard

[1] Isaiah xxxix. 2.

and saffron; calamus and cinnamon, with all trees of frankincense; myrrh, and aloes, with all the chief spices."

The last lines sum up the principal fragrant substances then in use, of which the following description may not be deemed out of place :—

Camphire is the same shrub which the Arabs call henna (*lawsonia inermis*), the leaves of which are still used by women in the East to impart a rosy tint to the palms of

Henna, or Camphire (*Lawsonia inermis*) with enlarged leaf and flower.

their hands and the soles of their feet. Its flowers are very fragrant, and are worn in chaplets round the neck, or used to decorate apartments and scent the air.

The true nature of Spikenard has been at all times the subject of much controversy. Ptolemy mentions it as an odoriferous plant, the best of which grew at Rangamati and on the borders of the country now called Bootan. Pliny says there are twelve varieties of it—

the best being the Indian, the next in quality the Syriac, then the Gallic, and, in the fourth place, that of Crete. He thus describes the Indian spikenard : " It is a shrub with a heavy thick root, but short, black, brittle, and yet unctuous as well; it has a musty smell, too, very much like that of the cyperus, with a sharp acrid taste, the leaves being small, and growing in tufts. The heads of the nard spread out into ears; hence it is that nard is so famous for its two-fold production, the spike or ear, and the leaf." [1] The price of genuine spikenard was then one hundred denarii per pound,[2] and all the other sorts, which were merely herbs, were infinitely cheaper, some being only worth three denarii per pound.

Galen and Dioscorides give a somewhat similar account of spikenard or *nardostachys*,[3] but the latter pretends that the so-called Syrian nard came in reality from India, whence it was brought to Syria for shipment. The ancients appear to have confounded spikenard with some of the fragrant grasses of India, which would account for the report that Alexander the Great when he invaded Gedrosia could smell from the back of his elephant the fragrance of the nard as it was trod upon by the horses' feet. This error was shared by Linnæus, who did not attempt to classify it, but was inclined to think it was the same as the *Andropogon nardus*, commonly called ginger-grass.

Sir William Jones, the learned orientalist, turned his

[1] Pliny's Nat. Hist. book xii. chap. 26.
[2] About £3 6s. 8d. of our money.   [3] From the Greek ναϱδοστάχυς.

serious attention to this question, and after a laborious investigation succeeded in establishing beyond doubt that the spikenard of the ancients was a plant of the valerianic order, called by the Arabs *sumbul*, which means " spike," and by the Hindús *jatamansi*, which signifies " locks of hair," both appellations being derived from its having a stem which somewhat resembles the tail of an ermine or of a small weasel. He consequently gave it the name of "Valeriana Jatamansi," under which it is now generally classed by botanists. It is found in the mountainous regions of India, principally in Bootan and Nepaul. Its name appears to be derived from the Tamil language, in which the syllable ந்ார் *nár* denotes any thing possessing fragrance, such as *nártum pillu*, " lemon-grass ;" *nárum panei*, " Indian jasmine ;" *nárta manum*, " wild orange," etc.

Spikenard.
(*Valeriana Jatamansi.*)

It is highly probable, however, that the word spikenard was often applied by the ancients as a generic name for every sort of perfume, as the Chinese now designate all their scents by the name of 香 *hëang*, which properly means *incense*, it being for them the type of all perfumes.

Saffron is composed of the dried stigmata of the flowers of the *crocus sativus*. Calamus is the root of the sweet flag (*crocus calamus*), and Cinnamon the bark of the *cinnamomum verum*.

Frankincense is an exudation from a sort of terebinth called *boswellia thurifera*, which is principally found in Yemen, a part of Arabia. In the time of Pliny it was only to be procured from that country, and he tells many marvellous stories respecting its mode of collec-tion and the difficul-ties in obtaining it. It has, however, since been discovered in some of the moun-tainous parts of India.

Myrrh is likewise an exudation from a tree called *balsamo-dendron myrrha*, found principally in Arabia and Abyssinia. The Greeks attributed a fabulous origin to this precious resin, hold-ing it to be produced by the tears of Myrrha,

Saffron (*Crocus Sativus*).

daughter of Cinyrus, king of Cyprus, who had been metamorphosed into a shrub. It is now scarcely used in perfumery, although it was such a favourite with the ancients.

The aloes mentioned here must not be confounded with the medicinal drug bearing the same name. It is the wood of a tree called *aloexylum agallochum* and is still greatly used in the East as a perfume principally for burning.

That these aromatics formed already an important branch of commerce, appears from the words used in the Canticles, "all powders of the merchant;" and it is

Frankincense (*Boswellia thurijera*).

equally evident they were applied to many purposes. Besides those that were burned, or used as perfumes, the allusion made to "the smell of the garments," shows that they laid them among their clothes, a custom also observed by the Greeks, as mentioned in Homer's "Odyssey," and kept up to the present day among Eastern nations. The most luxurious even applied scents to their couches, as we read in the Proverbs:—

The Aloes Tree.
(*Aloexylum Agallochum*).

"I have perfumed my bed with myrrh, aloes, and cinnamon." [1]

We cannot wonder that the Jews evinced such a taste for perfumes (a taste which they have retained to the present day), when we consider with what lavish hand Nature had showered her fragrant treasures upon them. Judea abounded with aromatic plants and shrubs, and well might Goldsmith hail it as a second Arabia:

> "Ye fields of Sharon, dress'd in flowery pride;
> Ye plains where Jordan rolls its glassy tide;
> Ye hills of Lebanon, with cedars crowned;
> Ye Gilead groves, that fling perfumes around;
> Those hills how sweet! those plains how wondrous fair!" [2]

The Egyptian custom of anointing the head of a guest to honour him was practised likewise by the Jews; thus when Jesus was sitting at table in Bethany,

[1] Proverbs vii. 17.   [2] "The Captivity."

in the house of Simon the leper, "there came a woman having an alabaster box of ointment of spikenard, very precious, and she brake the box and poured it on his head."[1]

The Jews had also borrowed from the Egyptians the practice of embalming their dead, for we see in the Gospel that after Jesus's death Nicodemus "brought a mixture of myrrh and aloes, about an hundred pound weight. Then took they the body of Jesus and wound it in linen clothes with the spices, as the manner of Jews is to bury."[2]

Soap does not appear to have been known by the Jews. It is true that the word *sope* occurs twice in the Bible,[3] but in these instances it may be permitted to doubt if it renders the true meaning of the Hebrew word בֹּרִית (*borith*). The Septuagint[4] translates it "herb," and the Latin Vulgate "the herb borith." Jarchi says it was an herb used by fullers for cleansing clothes, and Maimonides thinks it was the plant called by the Arabs *gazúl*, which, according to Jerome, grows abundantly in the moist parts of Palestine. Others again assert that it meant fuller's earth, or a saponaceous clay found in the east, which is still used there for the bath. Dr. Henderson in his new translation of Jeremiah, has it "potash,"[5] and he appears to be nearer the truth, for I strongly believe *borith* to have been nitrate of

---

[1] St. Mark xiv. 3.   [2] St. John xix. 39, 40.
[3] Jeremiah ii. 22; and Malachi, iii. 2.
[4] A Greek version of the Old Testament, supposed to be the work of seventy translators.
[5] Jeremiah and Lamentations, translated by Dr. Henderson, page 14.

potash, or common nitre. It may be objected that the words used by Jeremiah, " For though thou wash thee with nitre, and take thee much sope," show that nitre and borith were two different things. This I fully admit, but the substance called nitre by the ancients was in reality the natron of Egypt, a sesquicarbonate of soda which was found in several lakes in that country, and used for washing and also for embalming,[1] whilst our common nitre or saltpetre is a nitrate of potash. I am confirmed in this opinion, by the description of the holy incense found in the Talmud (Book Cheritoth,) which comprised בֹּרִית כַּרְשֶׁנָא, (*borith of Carshena*), probably a native nitre found at Carshena, and a very proper ingredient to promote combustion, if we admit it to be nitre, but difficult to explain if it is asserted to be a soap, a clay, or even an herb.

Jewish women were mostly endowed with great physical beauty—a gift which they have preserved to this time, throughout the work of ages, the changes of climes, and the innumerable hardships to which they have been submitted. Not contented, however, with their natural personal attractions, they tried to enhance them with various cosmetics, among which stood pre-eminent the Egyptian *kohl*, described in the last chapter. It was this artifice Jezebel resorted to when she was expecting Jehu; for, although the text says that she painted her *face*,[2] it was most probably her eyes to which she gave that dark hue which was thought so fascinating. Ezekiel explains this mode of painting more clearly

---

[1] See last chapter.   [2] 2 Kings ix. 30.

when he says, "Thou didst wash thyself, paintedst thine eyes, and deckedst thyself with ornaments."

The toilet implements used by the Jews were, like their perfumes, borrowed principally from their late masters, the Egyptians. They used the same sort of metal mirrors, and the brazen laver made by Moses for the tabernacle was composed of those belonging to the women of the congregation.

There is no country in the world where manners and

An Eastern Marriage Procession.

customs are so perpetuated from generation to generation as in the East. We find among the modern Arabs the same mode of life which was adopted by the patriarchs of old, and we may likewise form some idea of the costumes and habits of ancient Jewish women from those of the present occupants of the Holy Land. The marriage

An Eastern Bride.

procession represented on the previous page may give us some notion of the ancient way of performing that ceremony. The sweet aspersions and aromatic fumigations are still maintained; and in the annexed engraving of an Eastern bride, we recognise many of the ornaments, with the loss of which Isaiah threatens the daughters of Zion as a punishment for their wickedness :—

" In that day the Lord will take away the bravery of their tinkling ornaments about their feet, and their cauls, and their round tires like the moon,

" The chains, and the bracelets, and the mufflers,

" The bonnets, and the ornaments of the legs, and the head-bands, and the tablets, and the ear-rings,

" The rings and nose-jewels,

" The changeable suits of apparel, and the mantles, and the wimples, and the crisping-pins,

" The glasses, and the fine linen, and the hoods, and the veils.

"And it shall come to pass, that instead of sweet smell, there shall be a bad odour; and instead of a girdle, a rent; and instead of well-set hair, baldness; and instead of a stomacher, a girding of sackcloth; and burning instead of beauty." [1]

Of all the menaces held out by the prophet to the Hebrew women, that of baldness must have been the most severely felt by them, for they generally possessed very fine hair, which they wore confined in a net or caul, and ornamented with "round tires like the moon."

The men also kept their hair long, just as it grew; and Absalom's hair is said to have weighed two hundred shekels, which is about thirty-one ounces. Shorn locks were usually a sign of slavery; and

Jewish Captives at Babylon.

in this lamentable guise are represented Jewish captives at Babylon suing mercy from their conquerors. The priests had their hair cut every fortnight, while they were in waiting at the temple. The Nazarites, who made a vow of observing a more than ordinary degree of purity, were forbidden from touching their hair with a razor or scissors during its continuance, but when it ended they came to the door of the temple, and the priest shaved their heads, and burnt their hair on the altar.

[1] Isaiah iii. 18–24.

Josephus relates that, in grand ceremonies, king Solomon was preceded by forty pages, all scions of noble families, wearing their hair profusely powdered with gold-dust, which, glittering in the sun's rays, had a most brilliant effect. Our belles of the present time who patronise this mode of adornment and ascribe its invention to a modern illustrious lady, may not be aware that it is some three thousand years old, which confirms once more the truth of the adage, that "there is nothing new under the sun."

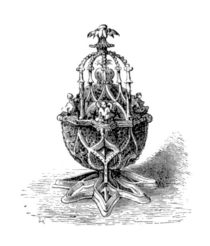

THE DEATH OF SARDANAPALUS.

CHAPTER IV.

THE ANCIENT ASIATIC NATIONS.

"In this pleasant soil
His far more pleasant garden God ordained,
Out of the fertile ground he caused to grow
All trees of noblest kind for sight, smell, taste."

MILTON'S PARADISE LOST.

HE strip of land running between those two mighty rivers, the Tigris and the Euphrates, which was called Mesopotamia by the ancients, and is named El Jezireh by its modern inhabitants, is supposed to have been the site of earthly Paradise. Some Scripture commentators, it is true, entertain the opinion that it was placed in Armenia; but out of the four rivers mentioned in Genesis as flowing through it, two being evidently the Tigris and the Euphrates, it seems more natural to suppose that Mesopotamia was the scene of that magnificent garden of Eden so beautifully described by Milton in his noble poem:—

> " It was a place
> Chosen by the immortal Planter, when he framed
> All things to man's delightful use: the roof
> Of thickest covert was inwoven shade,
> Laurel and myrtle, and what higher grew
> Of firm and fragrant leaf; on either side
> Acanthus, and each odorous bushy shrub,
> Fenced up the verdant wall; each beauteous flower,
> Iris all hues, roses and jessamine,
> Rear'd high their flourished heads between, and wrought
> Mosaic; under foot the violet,
> Crocus, and hyacinth, with rich inlay
> Broider'd the ground, more colour'd than with stone
> Of costliest emblem." [1]

That this favoured spot has preserved its natural beauties to the present day we may judge by Layard's description of the environs of the ancient city of Nimroud:—

"Flowers of every hue enamelled the meadows; not thinly scattered over the grass as in northern climes, but in such thick and gathering clusters, that the whole plain seemed a patchwork of many colours." [2]

Such an attractive region could not fail to be chosen by man at an early period for a dwelling-place; nor is it to be wondered at that it tempted more than once the ambitious invader to overrun its fertile plains and settle with his hordes in this desirable spot. It would be, however, quite out of my province to trace the his-

[1] " Paradise Lost," book iv. [2] "Nineveh and its Remains," vol. i. p. 78.

tory of the great Eastern empire from its foundation by Ashur, the son of Shem, and Nimrod, "the mighty hunter," to its conquest by Cyrus. I shall confine myself to what strictly appertains to my subject, and endeavour to delineate the manners and customs of the Assyrians, the Medes, the Persians, the Chaldeans, and other ancient Asiatic nations.

Besides the frequent reference to the Assyrians and Chaldeans which we find in the Bible, Herodotus, Xenophon, Diodorus Siculus, and other authors have transmitted to us, some curious and valuable information, respecting the mode of life of those luxurious people, which has been fully confirmed by modern discoveries.

For many centuries, Nineveh and Babylon, once the wonders of the universe, lived but in the memories of men. Their sites were scarcely known; and it was thought that every trace of them had disappeared from the face of the earth, when, some fifty years since, an English scholar and a French *savant*, Rich and Niebuhr, after long and patient researches, succeeded in lifting a corner of the shroud of sand and ruin which had so long covered the dead cities, and revealed to the astonishment of the world the splendours of Assyrian architecture. These pioneers of exploration were followed by Botta, Bonomi, Layard, and other ardent investigators, who, by dint of untiring perseverance and energy, rescued many valuable treasures from the mounds of rubbish which the present occupiers of the soil had allowed, in their careless ignorance, to accumulate over them. These interesting relics now enrich our museums; and in their graphic illustrations we may read, as in a written book, the manners and customs of a nation which rivalled Egypt in the arts of peace and war.

Baal, or Belus.

The Assyrians worshipped many deities, the principal of which were the sun, the moon, and the constellations. Baal, or Belus, the Egyptian Osiris, typified the sun, and was the most highly venerated of them all. Next came Astarte, or Mylitta, the Assyrian Venus, who, like Isis in Egypt, was honoured under the shape of the moon, which accounts for her being generally represented with a crescent on her head. Dagon, or the fish-god, was principally revered by the Phœnicians, to whom he was said to have taught the art of navigation.

Astarte, the Assyrian Venus.

On all the altars erected to these gods, incense and aromatic gums were burnt in great profusion, for we read in the Holy Writ of the "idolatrous priests that burn incense unto Baal, to the sun and the moon, and to the planets, and to all the host of heaven." [1]

Herodotus describes at great length the magnificent

Dagon, or the Fish-god.

Altar (Khorsabad).

temple erected in Babylon in honour of Baal, or Belus, which consisted in a series of eight huge towers raised one over the other, and thought by some to have been identical with the Tower of Babel. In the interior was a golden statue of the god, said to have weighed eight hundred talents (which made it worth about three millions of our money), and on the altar, which was also made of massive gold, they burned every year one thousand talents of pure incense. [2]

Nimrod's Statue and Altar.

Besides these deities the Assyrians also worshipped their ancient sovereigns, such

[1] 2 Kings xxiii. 5.　　　[2] Herodotus, book i.

as Nimrod, under whose statue an altar was found in one of the excavated monuments ; and Semiramis, their great queen, who had raised Babylon to its greatest state of splendour, and who was supposed to have been transformed into a dove, under which shape she was adored.

Altar on a High Place.

Their altars were not always placed in the temples ; they were sometimes raised on high places, a custom frequently alluded to in the Bible, and further illustrated by modern discoveries. The priests represented in the sculptures by the sides of the altar generally have in their hand a small square basket of wicker-work, the destination of which has greatly puzzled the *savans*. It may probably have been used to carry the aromatic gums and woods to be burned in the sacrifice. The consumption of these precious drugs was so large that, besides what the country produced, additional supplies were obtained from neighbouring nations. The Arabians alone, according to Herodotus, had to furnish a yearly tribute of one thousand talents of frankincense.

Assyrian Altar and Priests
(Khorsabad).

Zoroaster, during the reign of Darius Hystaspes, undertook to reform the religion of the Persians, and substituted the worship of fire for that of their various idols. Five times a day did his priests burn perfumes on the altar, and it was their duty to watch by turns so that the holy flame might not be extinguished.

The following origin was ascribed to the sacred fire: An astrologer once predicted at Babylon the birth of a child who would dethrone the king. The reigning monarch gave orders thereupon to have all women who were in a state of pregnancy put to death; but one of them, whose appearance had not betrayed her, gave secretly birth to the future prophet. The king having heard of it a short time after, sent for the child and tried to kill him with his own hand, but his arm was withered on the spot. He then had him placed on a lighted stake, but the burning pile changed into a bed of roses, on which the child quietly slept. Some persons present saved a portion of the fire, which was kept up to the present day, in memory of this great miracle. The king made two other attempts to destroy Zoroaster, but received punishment for his wickedness in the shape of a gnat which entered his ear and caused his death.[1]

Zoroaster's doctrines were adopted and upheld by the kings of the Sassanide dynasty, one of whom is represented on the accompanying medal, having on the obverse a pyreum, or holy altar-fire. When Persia was invaded by the Turks, his sectaries flew from the persecutions to which they were subjected by the Mahometans, and

[1] Tavernier, Voyage en Perse.

5

took refuge on the western coast of India, where they
continue to exercise their religion under the name of
Parsees, or Ghebers. They still keep up the sacred

Sassanide Medal.

fires on brazen altars, upon which they throw aromatic
gums in their ceremonies.

The luxurious and refined habits of the Assyrians in

Parsee Altar.

private life naturally in-
volved the use of per-
fumes and cosmetics.
Their last monarch, Sar-
danapalus, whom Col.
Rawlinson calls Assar-
adan-pal, carried this
passion to such an ex-
tent that he dressed and
painted like his women;
and, when driven to the last extremity by the rapid
advance of the conqueror, he chose a death worthy of
an Eastern voluptuary by causing a pile of fragrant
woods to be lighted, and, placing himself on it with
his wives and treasures, was sweetly suffocated by
aromatic smoke. Duris, however, and other historians
quoted by Athenæus, give another version of his death.

They say that Arbaces, one of his generals, having gone to visit Sardanapalus, found him painted with vermilion and clad in female garb. He was just in the act of pencilling his eyebrows when Arbaces entered, and the general was so indignant at the effeminacy of the monarch that he stabbed him on the spot.

Great as was the magnificence of Nineveh, it was scarcely equal to that of Babylon, which, according to ancient records, had a circumference of sixty miles, and contained the most gorgeous buildings and immense riches. Foremost among all these marvels were the celebrated hanging gardens which Nebuchadnezzar erected to please his wife Amytes, daughter of Astyanax, king of the Medes, and which were classed among the wonders of the world.

> " Within the walls was raised a lofty mound
> Where flowers and aromatic shrubs adorned
> The pensile garden.   For Nebassar's queen
> Fatigued with Babylonia's level plains,
> Sighed for her Median home, where Nature's hand
> Had scooped the vales and clothed the mountain side
> With many a verdant wood; nor long she pined
> Till that uxorious monarch called on Art
> To rival Nature's sweet variety.
> Forthwith two hundred thousand slaves uprear'd
> This hill—egregious work, rich fruits o'erhang
> The sloping vales, and odorous shrubs entwine
> Their undulating branches."

There by the side of the lofty cedar grew the mournful cypress and the elegant mimosa; but the favourite resort of the Queen was the bower where bloomed the rose and the lily, vying with each other in beauty and fragrance.

" And the jessamine faint, and the sweet tuberose,
  The sweetest flower for scent that blows,
  And all rare blossoms from every clime !"

We can easily conceive that people who professed such admiration for fragrant flowers had an adequate esteem for perfumes, and that when the season of the

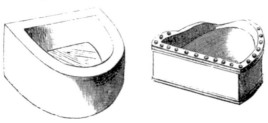

Assyrian Ointment Boxes

former was past they had recourse to the latter to per-petuate their enjoyment of " sweet smells." Babylon was, in fine, the chief mart for perfumes in the East, and Babylonian scents were celebrated far and wide.

Assyrian Perfume Bottles with cuneiform inscriptions (Nimroud).

The liquid essences were generally contained in bottles of glass or alabaster, and the ointment in boxes of porcelain or chalcedony. The bottles represented here with cuneiform inscriptions were found by Mr. Layard in the excavations at Nimroud.

The Babylonians themselves were great consumers of aromatics, for Herodotus tells us that they used to perfume their whole bodies with the costliest scents, and at their magnificent banquets fragrant cassolettes were kept constantly burning.

Cosmetics were also in much request among those luxurious people. Stibium, a preparation of antimony similar to the Egyptian kohl, they applied to the lids and corners of the eyes to make them appear larger and more brilliant. They used, besides, white and red paint for the face, and they rubbed their skin with pumice-stone to make it smooth.

Nicolaus of Damascus narrates the following curious anecdote, which illustrates the manners of the Babylonians. In the reign of Artæus, king of the Medes, one of his favourites, named Parsondes, a man renowned for his courage and strength, having observed that Nanarus, the governor of Babylon, was very effeminate in his person, shaving himself and using various cosmetics, he asked the king to transfer his post to him. Artæus refused, and Nanarus, having heard what had occurred, swore to be revenged on Parsondes. He caused him to be seized whilst he was hunting near Babylon, and having had him brought before him, inquired for what reason he had tried to supplant him. "Because," answered Parsondes, "I thought myself more worthy of the honour, for I am more manly and more useful to the king than you, who are shaven, and have your eyes underlined with stibium, and your face painted with white lead." Nanarus, on hearing this, delivered his enemy into the hands of a slave, to whom he gave strict orders to shave him, rub him with pumice-stone, bathe him twice a-day, anoint him, paint his eyes, and plait his hair like a woman's. This mode of treatment soon

Babylonian Banquet.

rendered Parsondes as effeminate as his rival; and, some time after, Artæus having sent one of his officers to Babylon to claim his favourite, Nanarus had him brought among one hundred and fifty female musicians before the ambassador, who could not recognise him, and took him for a woman.

The Medes were no less expert in the art of imparting artificial charms to their persons. Xenophon, in his Cyropedia,[1] relates that when Cyrus, at the age of twelve years, went with his mother to visit his grandfather, Astyages, King of the Medes, he found him adorned with paint round his eyes, colour on his face, and a magnificent wig of flowing ringlets. The boy, thinking all this was real, turned round to his mother, and exclaimed, in his naïve admiration, "Oh, mother, how handsome my grandfather is!"

The Persians borrowed from the Medes their taste for perfumes and cosmetics. Their kings usually spent their summers at Ecbatana, and their winters at Susa; the latter place was celebrated for its beautiful flowers, and especially the lily, which, being called *Souson* in the Persian language, gave its name to the town. Such was their predilection for perfumes that they usually wore on their heads crowns made, according to Dinon, of myrrh and a sweet-smelling plant called labyzus. In the palaces of monarchs and individuals of rank, aromatics were constantly burning in richly-wrought vessels, a custom of which we find an illustration in the annexed engraving, taken from the sculptures at Persepolis.

When Darius was vanquished by Alexander at the battle of Arbela, he left behind him in his tent, among other treasures, a casket filled with precious aromatics. Alexander, who at that time professed to despise such luxuries, had them thrown out, and replaced them with the works of Homer, who, by-the-bye, does not appear

[1] Xenophon, Cyrop. b. i. c. 3.

to have been so averse to sweet scents as his royal admirer, for he often praises them in his poems.   After the great conqueror had sojourned some time in Asia, he altered his views on the subject, for Athenæus tells us that he used to have the floor sprinkled with exquisite perfumes, and fragrant resins and myrrh were burnt before him, with other kind of incense.[1]

Incense-burning before a King. (Persepolis)

The greatest admirer of perfumes, however, among ancient Asiatic monarchs, seems to have been Antiochus Epiphanes, or the Illustrious,[2] King of Syria, who once

---

[1] Athenæus, Deipnosophists, b. xii.
[2] Athenæus calls him derisively Epimanes, or the Mad.

held some games at Daphne, where scents played a most important part.

In one of the processions that took place there were two hundred women sprinkling every one with perfumes out of golden watering-pots. In another, marched boys in purple tunics, bearing frankincense, and myrrh, and saffron, on golden dishes, and after them came two incense-burners made of ivy-wood, covered with gold, six cubits in height, and a large square golden altar in the middle of them. Every one who entered the gymnasium was anointed with some perfume contained in gold dishes. There were fifteen of these dishes, each holding different scents, such as saffron, cinnamon, spikenard, fenugreek, amaracus, lilies, etc. Thousands of guests were invited, and after being richly feasted were sent away with crowns of myrrh and frank-incense.

The same king was once bathing in the public baths, when some private person, attracted by the fragrant odour which he shed around him, accosted him, saying, "You are a happy man, O king: you smell in a most costly manner." Antiochus, being much pleased with the remark, replied, "I will give you as much as you can desire of this perfume." The king then ordered a large ewer of thick unguent to be poured over his head, and a multitude of poor people soon collected around him to gather what was spilled. This caused the king infinite amusement, but it made the place so greasy that he slipped and fell on his royal back in a most undignified manner, which put an end to his merriment.

All other Asiatic nations made great use of perfumes, and paid great attention to their toilet ; but none, perhaps, exceeded in that taste the Lydians, who were most effeminate, and whom Xenophanes describes as

> " Boasting of hair luxuriously dress'd,
> Dripping with costly and sweet-smelling oils.

The Egyptian custom of embalming does not appear to have been practised in the same manner by the Assyrians or Babylonians. Herodotus says that the latter preserved the bodies of their dead with honey, but this would not have been sufficient without the admixture of some aromatic substances. M. Botta found a great number of funereal urns at Nineveh, which only contained fragments of bones, the bodies having been transformed into clay.

No ancient nation devoted such care to the hair and beard as the Assyrians. The mass of luxuriant curls falling over the shoulders and the elaborately plaited beard are so familiar to those who have visited our museums that I need not give any enlarged description

of this fashion. The kings usually had gold thread interwoven with their beard, which, contrasting with its dark hue, had a most brilliant effect. Their head-dress was of a semi-conical form, and enriched with pearls and jewels. Cyrus is said to have been the first to wear the tiara, but he is represented on a monument at

King's Head-dress. Persepolis with a most peculiar head-dress, which, if ornamental, must have been somewhat

inconvenient, as the reader may judge from the annexed
engraving, which would not form a bad
design for a candelabrum.

Ladies wore their hair flowing in
long ringlets over their shoulders, and
simply confined by a band round the
head, as shown in the accompanying
illustration. They wore massive ear-
rings and a profusion of jewels, and
were mostly pretty. Those, however,

Cyrus' Head-dress.
(Persepolis.)

who had not been favoured with Nature's gifts did not

on that account remain single; for, by a very curious
custom established at Ba-
bylon, all marriageable
girls were assembled to-
gether at a certain time,

Assyrian Ear-rings.

and the rich suitors selected first the handsomest brides,
and paid down a dowry,
which was given to the
others, who by means of
this easily found husbands
among the young men who
cared more for money than
beauty.

All the Asiatic people attached the greatest value
to their hair; and well did
Mausolus, king of Caria,

Babylonian Ladies.

turn this fondness to account when he resorted to the

following stratagem to replenish his impoverished exchequer. Having first had a quantity of wigs manufactured and carefully stored in the royal warehouses, he published an edict compelling all his subjects to have their heads shaved. The unfortunates had to submit; and when, a few days after, the monarch's agents went round offering them the perukes destined to cover their denuded polls they were glad to buy them at any price. No wonder that Artemisia could not console herself for the loss of such a clever husband; and that, not satisfied with drinking his ashes every day mixed with her wine, she exhausted the treasures of the state in erecting to his manes a splendid monument, which was reckoned one of the wonders of the world.

VENUS TOILET.

CHAPTER V.

THE GREEKS.

Ἀμβροσίη μὲν πρῶτον ἀπὸ χροὸς ἱμερόεντος
Λύματα πάντα κάθηρεν, ἀλείψατο δε λίπ' ἐλαίῳ
Ἀμβροσίῳ, ἑδανῷ, τό ῥά οἱ τεθυωμένον ἦεν.   HOMER.

NUMEROUS as the stars were the deities adored by the ancient Greeks. In Hesiod's time they had already attained the respectable number of thirty thousand,[1] and new ones were being constantly manufactured or adopted from other nations. There were many different rites observed in their worship, but they nearly all comprised sacrifices, which were offered not only in the temples but also in private houses, where altars were erected for

[1] Hesiod, Oper. i. 250.

that purpose.   No Greek commenced a journey or any other enterprise of greater or lesser moment without having first sought to propitiate the god whose protection he thought he might require in his undertaking, by sacrificing the animal consecrated to that particular deity. Thus an ox was offered to Jupiter, a dog to Hecate, a dove to Venus, a sow to Ceres, and a fish to Neptune.   The victim was laid on the altar decked with garlands of fragrant herbs or flowers, and

Private Altar.

Patera.

burned with frankincense, accompanied with libations of wine out of a flat vessel called patera. This formed the complete oblation described by Hesiod.

> " Let the rich fumes of od'rous incense fly,
>   A grateful savour to the powers on high ;
>   The due libation nor neglect to pay,
>   When evening closes, or when dawns the day." [1]

In the more ordinary sorts of sacrifices, incense alone was burned on the *thytérion*, or incense altar, as represented in the accompanying engraving. At all the numerous religious festivals held in Greece, aromatics were consumed in large quantities. The principal of these *fêtes* were the Panathenæa, in honour of Minerva; the Eleutheria, celebrated at Platæa, in the temple of Jupiter; and the Dyonisia, of which Bacchus was the

[1] Hesiod, Oper. i. 334.

hero; but none equalled in magnificence the Eleusinian mysteries, instituted in honour of Ceres. The latter festival lasted nine days, during which the mystæ, or initiates, were gradually subjected to a series of terrifying trials to test their fortitude. Those who had succeeded in braving the most hideous apparitions, the most ferocious monsters, and the most appalling dangers, were intro-duced on the ninth day into the

Incense Altar.

temple of the goddess, where her statue, covered with gold and precious stones, shone amidst a thousand lights. The altar, smoking with the purest incense,

Greek Altar.

was surrounded by a crowd of priests clad in purple, and crowned with myrtle; and above them, on a splendid throne, sat the Hierophant, or high priest, who expounded to the adepts the mysteries of the goddess, and described to them the joys which awaited them in return for their cour-age. In the midst of the Elysian fields they were to find a golden city with emerald ramparts, ivory pavement, and cinnamon gates. Around the walls flowed a river of perfumes one hundred cubits in width, and deep enough to swim in. From this river rose an odorous mist, which enveloped the whole place and shed a refreshing and fragrant dew. There were to be, besides, in this fortunate city, three hundred and sixty-five fountains of honey and five hundred of the sweetest essences. This description, taken from a Greek author, bears a singular resemblance to that of the marvels of Mahomet's paradise, promised to the Mussulmans in the Koran, as will be seen hereafter, and shows the passionate fondness of both people for perfumes.

The Greeks, with their lively imagination, constantly mixing up fable with reality, ascribed a divine origin to perfumes, which they numbered among the attributes of their deities. Thus, as I have remarked before in the first chapter, the early poets never mention the apparition of a goddess without speaking of the ambrosial fragrance which she shed around her. The gods who revelled in nectar and ambrosia, food unknown to mortals, indulged also in delicious perfumes specially reserved to their use. Homer thus describes Juno's toilet operations when she repairs to her bower before meeting Venus :—

> " Here first she bathes, and round her body pours
> Soft oils of fragrance, and ambrosial showers.
> The winds, perfumed, the balmy gale conveys
> Through heaven, through earth, and all th' aërial ways.
> Spirit divine ! whose exhalation greets
> The sense of gods with more than mortal sweets." [1]

Sometimes good-natured deities condescended to bestow some of these exquisite aromatics upon their own

---

[1] Iliad, xiv.

protégés as a mark of special favour. Thus, when Penelope prepares to receive her suitors, Eurynome advises her to dispel her grief, and diffuse " the grace of unction over her cheeks." The virtuous matron refuses in the following terms :—

> " Persuade not me, though studious of my good,
> Eurynome ! to bathe or to anoint
> My face with oil; for when Ulysses sail'd,
> On that same day, the Pow'rs of Heav'n deform'd
> And wither'd all my beauties."

Pallas, however, visits her during her slumbers, and sheds over her some wonderful perfume, which was probably called in those times " The Venus Bouquet."

> " The glorious goddess clothed her as she lay
> With beauty of the skies ; her lovely face
> With such ambrosial unguent first she bathed
> As Cytherea, chaplet-crowned, employs
> Herself, when in the sight-entangling dance
> She joins the Graces." [1]

Phaon, the Lesbian pilot, having once conveyed in his vessel to Cyprus a mysterious passenger, whom he discovers to be Venus, receives from the goddess, as a parting gift, a divine essence, which changes his coarse face into the most beautiful features. Poor Sappho, who sees him after his transformation, becomes smitten with his charms, but, finding her love unrequited, is driven to seek a watery grave. This miracle certainly beats all the vaunted achievements of modern perfumery, even including the " patent enamelling process," which, if applied to gentlemen, would not, I am afraid, attract many " Sapphos."

The persons skilled in preparing perfumes—and they

[1] Odyssey, xviii.

6

were mostly women—were deemed by the Greeks, with their love of the marvellous, to be magicians. Thus we have Circe detaining Ulysses in her isle by means of spells, which were chiefly sweet fumigations; and Medea boiling old Eson in an aromatic bath, and turning him out a perfect juvenile—an operation, by-the-bye, which few of our old beaux would submit to, whatever may be their wish to become young again.

The nymph Œnone was supposed to have imparted to Paris some of the secrets of Venus's toilet, and it was by means of these cosmetics that the fair Helen acquired that transcendent beauty which was so fatal to both Greeks and Trojans. These secrets she revealed to her countrywomen on her return from Troy, and thus we have the perfection of Greek perfumery accounted for.

In those ancient times, besides the fragrant gums burned as sacrifices, the only perfumes known appear to have been in the shape of oils scented with flowers, and principally the rose. Homer generally designates them under the name of ἔλαιον, claion (oil), adding sometimes the epithet of "rosy" or "ambrosial." At a later period the Ionians introduced a greater variety of essences chiefly borrowed from Asiatic nations, who were then more versed in the art.

Their use became so prevalent at one time, that Solon issued an edict prohibiting the sale of perfumes; but, like all sumptuary laws, it was "more honoured in the breach than the observance," for perfumers' shops still continued to be the resort of loungers, as modern *cafés* are in the south of Europe. Even the tattered cynic,

Diogenes, did not disdain to enter them now and then, leaving his tub at the door; but, with a praiseworthy spirit of economy, he always applied the ointments he bought to his *feet;* for, as he justly observed to the young sparks who were mocking him for his eccentricity, "when you anoint your head with perfume it flies away into the air, and the birds only get the benefit of it; whilst if I rub it on my lower limbs it envelopes my whole body, and gratefully ascends to my nose."

The general name for perfumes was μύρον (*myron*), which, according to Chrysippus, was derived from the word *moron* (trouble), "owing to the vain and unprofitable labour of compounding it." I put this down, however, as the detestable pun of a man who had "no perfume in his soul," and am more inclined to believe it came from the word myrrh, as being the best known of aromatics.

The Greeks in the time of their splendour used a great variety of scents and unguents, the principal of which are thus described at full length by Apollonius of Herophila, in his "Treatise on Perfumes," quoted by Athenæus:[1]—

Alabaster Scent-bottle.

"The iris is best in Elis and at Cyzicus; the perfume made from roses is most excellent at Phaselis, and that made at Naples and Capua is also very fine. That made from crocus (saffron) is in the highest perfection at Soli in Cilicia, and at Rhodes. The essence of spikenard is best at Tarsus, and the extract of vineleaves is made best at Cyprus and at Adramyttium.

[1] Deipnosophists, b. xv. c. 38.

The best perfume from marjoram and from apples comes from Cos. Egypt bears the palm for its essence of Cypirus, and the next best is the Cyprian and Phœnician, and after them comes the Sidonian. The perfume called Panathenaicum is made at Athens, and those called Metopian and Mendesian are prepared with the greatest skill in Egypt. But the Metopian is made from oil which is extracted from bitter almonds. Still, the superior excellence of each perfume is owing to the purveyors, and the materials, and the artist, and not to the place itself, for Ephesus formerly, as men say, had a high reputation for the excellence of its perfumery, and especially of its megallium, but now it has none. At one time, too, the unguents made in Alexandria were brought to high perfection on account of the wealth of the city and the attention that Arsinoe and Berenice paid to such matters ; and the finest extract of roses in the world was made at Cyrene while the great Berenice was alive. Again, in ancient times the extract of vine-leaves made at Adramyttium was but poor ; but afterwards it became first-rate, owing to Stratonice, the wife of Eumenes. Formerly, too, Syria used to make every sort of unguent admirably, especially that extracted from fenugreek, but the case is quite altered now. And long ago there used to be a most delicious unguent extracted from frankincense at Pergamus, owing to the invention of a certain perfumer of that city, for no one else had ever made it before him ; but now none is made there."

Theophrastus also wrote a book on scents, in which he says that some perfumes are made of flowers, as, for instance, from roses, white violets, and lilies—some from stalks or leaves, and some from roots.

The name of the perfumes generally indicated the ingredients from which they were prepared, but others were called after their inventor. Thus the Megallium was made by a perfumer named Megallus—

> "And say you are bringing her such unguents
> As old Megallus never did compound."[1]

Peron was also a celebrated Athenian perfumer, often quoted by ancient authors:—

> "I left the man in Peron's shop just now
> Dealing for ointment; when he has agreed,
> He'll bring you cinnamon and spikenard essence."[2]

Baccaris and Psagdas, or Psagdès, were two perfumes much in vogue:—

> "I then my nose with baccaris anointed,
> Redolent of crocus."[3]

> "Come, let me see what unguent I can give you:
> Do you like psagdès?"[4]

> "She thrice anointed with Egyptian psagdas."[5]

The most luxurious applied a different perfume to each part of their body, as we find in Antiphanes:—

> "He really bathes
> In a large gilded tub, and steeps his feet
> And legs in rich Egyptian unguents;
> His jaws and breasts he rubs with thick palm oil,
> And both his arms with extract sweet of mint;
> His eyebrows and his hair with marjoram,
> His knees and neck with essence of ground thyme."

The greatest consumption of aromatics, however, took place in their entertainments. Already in Homeric times it was customary to offer to the guests a bath

[1] Strattis, Medea.   [2] Antiphanes, Antea.   [3] Hipponax.
[4] Aristoph. Daitaleis.   [5] Eubulus.

followed by sweet unctions before sitting to table. Thus when Telemachus and Pisistratus are received by Menelaus they descend to the baths—

> "Where a bright damsel train attends the guests
> With liquid odours and embroider'd vests ;
> Refreshed they wait them to the bower of state,
> Where circled with his peers Atrides sate." [1]

At a later period perfumes were not only used for ablutions prior to the entertainment, but were also brought in, during the feast, in alabaster or gold bottles, with flower garlands to crown the guests.[2] Philoxenus, in his play called "The Banquet," says—

> "And then the slaves brought water for the hands,
> And soap[3] well mix'd with oily juice of lilies,
> And pour'd o'er the hands as much warm water
> As the guest wish'd. And then they gave them towels
> Of finest linen, beautifully wrought,
> And fragrant ointments of ambrosial smell,
> And garlands of the flow'ring violet."

Xenophanes gives a still more ample description of a Grecian entertainment :—

> "The ground is swept, and the triclinium clear,
> The hands are purified, the goblets, too,
> Well rinsed ; each guest upon his forehead bears
> A wreath'd flow'ry crown ; from slender vase
> A willing youth presents to each in turn
> A sweet and costly perfume ; while the bowl,
> Emblem of joy and social mirth, stands by,
> Fill'd to the brim ; another pours out wine
> Of most delicious flavour, breathing round
> Fragrance of flowers, and honey newly made,
> So grateful to the sense, that none refuse ;
> While odoriferous gums fill all the room.

---

[1] Odyssey, iv.   [2] Athenæus, Deipnos., b. xv., c. 36.

[3] Although the original Greek word σμῆγμα (smègma) is usually translated "soap," I believe it only meant a kind of scented clay, still used in the East. for the Greeks were unacquainted with soap.

Water is served, too, cold, and fresh, and clear;
Bread, saffron tinged, that looks like beams of gold.
The board is gaily spread with honey pure
And savoury cheese.   The altar, too, which stands
Full in the centre, crown'd with flow'ry wreaths;
The house resounds with music and with song."

Although the preceding details indicate a high state of luxury in Grecian entertainments, some voluptuaries were not even satisfied with those means of enjoyment, but sought to increase them by resorting to all sorts of ingenious devices, such as that mentioned in the "Settler of Alexis":—

" Nor fell
His perfumes from a box of alabaster;
That were too trite a fancy, and had savour'd
O' the elder time—but ever and anon
He slipp'd four doves, whose wings were saturate
With scents, all different in kind—each bird
Bearing its own appropriate sweets—these doves
Wheeling in circles round, let fall upon us
A shower of sweet perfumery, drenching, bathing
Both clothes and furniture, and lordlings all.
I deprecate your envy when I add
That on myself fell floods of violet odours."

This mode of using perfumes during their banquets was not only adopted on account of the pleasure they created, but because a beneficial effect was ascribed to them, especially when rubbed on the head :—

" The best recipe for health
Is to apply sweet scents unto the brain."

Anacreon also recommends the breast to be anointed with unguents, as being the seat of the heart, and considering it an admitted point that it was soothed by fragrant smells.   Another virtue the Greeks attributed to perfumes, and not the least in the sight of the

Epicureans, was, that it enabled them to drink more wine without feeling any ill effects from it. This belief, however justified it might have been, is alluded to by many authors. The most refined votaries of Bacchus were not satisfied with the *external* use of aromatics : they also applied them to improve the taste of their wine. Some of these were prepared with odorous resins, such as the myrrhine, which was flavoured with myrrh ; others had simple honey or fragrant flowers infused in them.

If scents were in favour with the wealthy and luxurious Athenians, they were not so with the philosophers, who condemned their use as effeminate. Xenophon relates that Socrates, being once entertained by Callias, was offered some perfumes, but he declined them, saying they were only fit for women, and that for men he preferred the smell of the oil used in the gymnasia. " For," added he, "if a slave and a freeman be anointed with perfumes, they both smell alike; but the smell derived from free labours and manly exercise ought to be the characteristic of the freeman." I am bound to add, as a faithful historian, even at the risk of damaging Socrates in the eyes of my fair readers, that he equally disapproved of baths, considering cleanliness no essential part of wisdom.

Although the elaborate Egyptian system of bathing had been partly adopted by the Greeks, they never gave it that development which it acquired afterwards with the Romans. They were generally satisfied with more limited ablutions, performed in a marble basin situated in some public place, whilst the ladies attended at home to the duties of the toilet. The engravings given here are taken from antique sculptures, or from specimens in the British Museum.

Perfumes, as I have said before, were generally supposed to possess medicinal properties, and the recipes of the most celebrated essences and cosmetics were inscribed on marble tablets both in the temples of Esculapius and of Venus. The priestesses of various deities succeeded the ancient magicians, and dispensed their preparations, which were supposed to be endowed with par-

Public Washing Basin.

Ladies' Toilet Basin.

ticular virtues, and competed successfully for a long time with the less divine productions of ordinary perfumers.

Milto, a fair young maiden, the daughter of an humble artisan, was in the habit of depositing every morning garlands of fresh flowers in the temple of Venus, her poverty preventing her from indulging in richer offerings. Her splendid beauty was once nearly destroyed by a tumour which grew on her chin, but she saw in a dream the goddess, who told her to apply to it some of the roses from her altar. She did so, and recovered her charms so completely that she eventually sat on the Persian throne as the favourite wife of Cyrus.

Since that time the reputation of the rose was established as a flower no less beneficial than beautiful, and it formed the basis of many lotions, both useful and ornamental, for as Anacreon says—

> " The rose distils a healing balm,
>   The beating pulse of pain to calm."

Even to the present day the queen of flowers has preserved

Greek Ladies at their Toilet.

its double fame, and is to be found equally on the shelves of the apothecary and in the laboratory of the perfumer.

Greek Girl Painting.

All the Grecian cosmetics, however, were not so innocent as the rose. The sedentary life of women deprived them of a great part of their natural freshness and beauty, and they sought to repair their loss by artificial means. They painted their face with white lead, and their cheeks and lips with vermilion or a root called pœderos, which was similar to alkanet-root. This was applied with the finger, or with a small brush, as represented in the annexed engraving, taken from an antique gem. They also used Egyptian kohl, for darkening the eyebrows and eyelids, and various other preparations for the complexion, which will be more amply described in the next chapter, as they were nearly all afterwards adopted by the Romans.

Hair dye was often employed by those who wished to emulate old Eson's renovation without having recourse to the boiling process. Laïs, who was as celebrated for her wit as for her beauty, having once repulsed the sculptor Miron, who at the age of seventy fell desperately in love with her, the discomfited suitor attributed his rejection to his white locks; he therefore had them dyed of a splendid black colour, and returned the next day hoping for better success. But he was doomed to disappointment, for Laïs replied, laughing, to his demands, "How can I grant thee to-day what I refused to thy father yesterday?"

From the earliest times perfumes were used by the Greeks in their funeral rites. Homer represents Achilles with his attendants paying thus the last honours to his friend Patroclus :—

> "The body then they bathe with precious toil,
> Embalm the wounds, anoint the limbs with oil," [1]

Even to an enemy it was considered a duty to pay this last tribute; and we find Achilles having the body of Hector anointed and perfumed before he returns it to Priam.[2]

---

[1] Iliad, xvii.　　　　[2] Iliad, xxiv.

A pile was usually raised to burn the bodies of the dead, and the friends of the deceased stood by during the operation, throwing incense on the fire, and pouring libations of wine. The bones and ashes were afterwards collected, washed with wine, and, after mixing them with precious ointments, inclosed in funereal urns, such as the annexed specimens taken from the British Museum. Agamemnon is described by Homer in the "Odyssey," informing Achilles how this ceremony had been performed upon him :—

Funereal Urns.

> " But when the flames your body had consumed,
> With oils and odours we your bones perfumed,
> And wash'd with unmix'd wine." [1]

It was also customary to strew fragrant flowers and shed sweet perfumes over the tombs of the dead ; and Alexander is said to have paid this mark of respect to Achilles, whose monument he anointed and crowned with garlands when he visited Troy.

Perfumes were thought such an essential part of funeral ceremonies that scent-bottles were painted on the coffins of the poorer class of people as a sort of empty consolation for the absence of the genuine article.[2]

Anacreon, as a true voluptuary, preferred enjoying

[1] Odyssey, xxiv.   [2] Aristophanes, Eccles.

perfumes and flowers in his lifetime to having them
offered to his manes after his death. He exclaims in
one of his odes—

> " Why do we shed the rose's bloom
>   Upon the cold insensate tomb?
>   Can flowery breeze or odour's breath
>   Affect the slumbering chill of death?
>   No, no; I ask no balm to steep
>   With fragrant tears my bed of sleep;
>   But now while every pulse is glowing,
>   Now let me breathe the balsam flowing;
>   Now let the rose, with blush of fire,
>   Upon my brow its scent expire." [1]

The cares and duties of the *toilette* were deemed of
such importance that a tribunal was instituted at Athens
to decide on all matters of dress, and a woman whose
peplon, or mantle, was not of correct cut, or whose
head-dress was neglected, was liable to a fine, which
varied according to the gravity of the offence, and
sometimes reached the high sum of a thousand
drachmæ. I must say, however, that Grecian ladies
do not seem to have required such a law to make them
study their personal appearance; their own coquetry
acted, no doubt, as a still more powerful stimulant,
and the antique specimens we have left would tend to
show that they possessed excellent taste, especially in
their modes of dressing the hair.

In ancient times the hair of both sexes was rolled up
into a kind of knot on the crown of the head, which
mode was called *krobylos* for the men, and *korymbos* for
the women. The greatest luxury of the latter at that

---

[1] Anacreon, Ode xxxii.

period was to ornament that knot with a golden clasp
in the shape of a grasshopper. This simple ornament
was however discarded in later times, and many different fashions were adopted, among which the most pre-

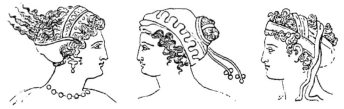

Mitra Head-dresses.

valent were the *kekryphalos*, the *sakkos*, and the *mitra*.
The first was a caul of network, which we have already
found among the Jews, and which we shall find again
in many other epochs and nations; a fact which, by-
the-bye, somewhat impairs its claims to novelty put
forth a very short time since. The sakkos was a close

Sakkos Head-dress.   Korymbos Head-dress.

bag, made generally of silk or wool; and the mitra,
which was of Asiatic origin, was a band of cloth dyed
of the richest colours and bound in various ways round
the head. There were many other modes of wearing
the hair, such as the *strophos*, the *nimbo*, the *kredemnon*,
the *tholia*, etc., of which the annexed illustrations will

convey a better idea than a written description, and my fair readers will no doubt find among them some which

Strophos Head-dresses.

would be almost *à la mode* in a drawing-room of the present day.

The men used to cut off their hair when they attained

Nimbo Head-dresses.

the age of puberty, and dedicate it to some deity. Theseus is said to have repaired to Delphi to perform this

Kredemnon Head-dress.   Tholia Head-dress.

ceremony, and to have consecrated his shorn locks to Apollo.

After this they allowed their hair to grow long again, and only cut it off as a sign of mourning. Thus, at the funeral of Patroclus, the friends of Achilles cut off their hair, and

"O'er the corse their scatter'd locks they threw." [1]

In some parts of Greece, however, where it was customary to wear the hair short, they allowed it to grow long when in mourning—

"Neglected hair shall now luxurious grow,
And by its length their bitter passion show." [2]

Another striking proof that external marks of grief are only matters of convention, and that the white garb of the Chinese mourner may be coupled with as much real sorrow as our sable habiliments.

[1] Iliad, xxiii.   [2] Cassandr. 973.

A ROMAN LADY'S BOUDOIR.

CHAPTER VI.

The Romans.

Discite, quæ faciem commendet cura, puellæ,
Et quo sit vobis forma tuenda modo.   Ovid.

OME, during the first period of its history, knew but little of the luxuries of civilisation. Its inhabitants, constantly at war with their neigh-bours, cared not for the arts of peace ; and their unshorn locks and shaggy beards were more calculated to strike terror into their enemies than to captivate the eyes of the fair sex. The only perfume they indulged in at that time was perhaps a bunch of verbena or other fragrant plant, which they plucked in the fields and hung over their door to keep away the evil eye, *il malocchio*, still so dreaded by their modern descendants. Even their gods did not then fare much better, and the sacrifices offered to them were, as Ovid says, of the plainest description:[1]

> " In former times the gods were cheaply pleased,
> A little corn and salt their wrath appeased,
> Ere stranger ships had brought from distant shores
> Of spicy trees the aromatic stores ;
> From India or Euphrates had not come
> The fragrant incense or the costly gum ;
> The simple savin on the altars smoked,
> A laurel sprig the easy gods invoked,
> And rich was he whose votive wreath possess'd
> The lovely violet with sweet wild flowers dress'd."

As, however, the Romans extended their conquests towards the provinces of Southern Italy colonized by the Greeks, which had received the name of Magna Græcia, they gradually adopted the manners of the countries they had vanquished, and became initiated in all the refinements of luxury. They imitated, likewise,

Incense Altar.
(Ara Turicrema.)

their religious ceremonies ; and in the various implements and paintings found at Herculaneum and Pompeii, the Grecian origin is easily discernible. To describe the Roman modes of worship would, therefore, be a repetition of the last chapter : we should find precisely the same things under different names. Thus the incense casket used for sacrifices, and called by the Greeks λιβανωτρίς (*libanótris*), became the "acerra;"

[1] Fastor. iii. 337.

the θυτήριον (*thytérion*), or altar, was changed into "ara turicrema ;" and the Grecian θυμιατήριον (*thumiatérion*) became the Roman "turibulum."

The accompanying illustrations will give some idea

Incense Casket.
(Acerra.)

Censer (Turibulum).

of the usual forms of these various implements. The incense casket is taken from a basso-relievo in the Capitol Museum, the altars from ancient paintings, and the censer from an original in bronze found at Pompeii. The chariot represented on the next page was also discovered in some excavations, and was used in the temples to carry incense to the various altars.

Roman Altar.

Funeral rites are so much grafted on religious ideas, that we must naturally expect to find the same resemblance between the Greek and the Roman ceremonies. In the early times of Rome, the dead were buried; but when Greek manners were adopted, they were burnt in the way already described, and the bones gathered in a funereal urn, with perfumes more or less costly, according to the fortune of the deceased, or the extent of gratitude of his heirs. Rich people usually had sepulchral chambers built, like the one represented

here, where they placed the funereal urns of all members of their family.

Although in private life Greek customs were likewise imitated, those of the Romans assumed peculiar features which it may be

Sepulchral Chamber.

interesting to study. A Sicilian named Ticinus Menas, in the year 454, brought into Rome the mode of shaving the beard, and sent to his country for a troop of clever barbers, who established their shops under the porticos of Minucius, near the temple of Hercules. Scipio Africanus and the *élite* of the patricians adopted the new fashion, and in a short time smooth chins, and hair redolent with ointments, became the rage, beards being left to slaves and common people.

The use of perfumes in Rome may be dated from that period, and became soon so prevalent, that Lucius Plotius, being proscribed by the tri-

Incense Chariot.

umvirs, and having taken refuge at Salernum, was betrayed in his hiding-place by the smell of his unguents, and put to death. After the defeat of Antiochus and the conquest of Asia, the abuse became still greater; and in the year 565, wishing to put a stop to it, the consuls, Licinus Crassus and Julius Cæsar, published a law forbidding the sale of "exotics," meaning thereby all sorts of perfumes which then came from abroad. This edict, however, was no better observed than Solon's had been at Athens, and did not in any way diminish the consumption of aromatics, which reached its greatest height under the reign of the emperors.

Among the latter, Otho was one of the most ardent votaries of the perfumer's art, for Suetonius[1] tells us that, even when going on a military campaign, he carried with him a complete arsenal of essences and cosmetics to adorn his person and preserve his complexion. Juvenal, in one of his satires, thus ridicules him for his effeminacy :—

> " Oh! noble subject for new annals fit,
> In musty Fame's report unmentioned yet,
> A looking-glass must load th' Imperial car,
> The most important carriage of the war;
> Galba to kill he thought a general's part;
> But as a courtier used the nicest art
> To keep his skin from tan; before the fight
> Would paint and see his soil'd complexion right." [2]

Caligula spent enormous sums for perfumes, and plunged his body, enervated by excesses, in odoriferous baths.[3] Nero was also a great admirer of sweet scents; and at Poppæa's funeral he consumed more incense than Arabia could produce in ten years. In his golden palace the dining-rooms were lined with movable ivory

---

[1] Sueton., b. viii.   [2] Juvenal, Sat. i.   [3] Suet., b. iv.

plates, concealing silver pipes, which were made to throw on the guests a sweet rain of odoriferous essences.[1]

The Romans had borrowed from the Egyptians the use of the public bath, to which they resorted almost daily—a very necessary measure to insure health and cleanliness, when we consider that they wore neither linen nor stockings. Their baths, or *thermæ*, were very

Roman Baths.

magnificent buildings, as we may judge from the ruins still extant. The principal establishments of that kind had been built at various times by the Emperors, and bore their names. The largest were those of Agrippa, Nero, Titus, Domitian, Antoninus, Caracalla, and Diocletian. They were open to the public at first on the payment of a *quadrans*, or a little less than a farthing of our money. Agrippa bequeathed his garden and baths

[1] Suet., b. vi.

to the Roman people, and assigned particular estates to
their support, that they might enjoy them gratuitously.
The plan of those baths was so well devised that it
deserves a particular description. On entering them,
the bathers first proceeded to undress, and gave their
clothes to guard to persons called *capsarii*, who were
hired for the purpose. They went then into the
unctuarium, or *eleothesium*—a room marked at the back
of our engraving—where all the perfumes and ointments were kept in large jars, making it somewhat
resemble a modern apothecary's shop. There they received a preliminary unction of cheap oils, and next
proceeded to the *frigidarium*, or cold bath, where they
went through the first course of ablution. Thence they
passed into the *tepidarium*, or tepid bath, and after that
they entered the *caldarium*, or hot bath, where
the temperature was maintained at a high degree by means of a furnace placed underneath,
called *hypocaustum*. There, whilst undergoing
profuse perspiration, they scrubbed their
skin with a sort of bronze curry-comb called
strigil—somewhat in the same fashion as modern grooms treat their horses—and dropped
on their body at the same time a little scented
oil out of a small bottle named *ampulla*. Those

Strigil and
Ampulla.

who could afford it had this operation performed upon
them by the bath attendants, called *aliptes*, or by their
own slaves, whom they brought with them for that
purpose.

There is a story told of the Emperor Hadrian, who,

one day bathing with the common people, and seeing an old soldier, whom he had known among the Roman troops, rubbing his back against the marble wall, asked him why he did so. The veteran answered that he had no slave to attend on him. Whereupon the Emperor presented him with two slaves and enough money to maintain them. A few days afterwards, two old men, enticed by the good fortune of the veteran, began to rub

Tepidarium at Pompeii.

themselves also against the wall, in the hope of attracting the Emperor's attention; whereupon Hadrian, perceiving their drift, told them that if they had no slaves they had better rub their backs against each other.

The accompanying engraving represents the *tepidarium* of the baths at Pompeii, with the three bronze benches on the sides, and the stone at the end, such as they were actually found. The compartments above were probably used to keep unguents and perfumes, and it is supposed that, as these baths were of small dimensions, this room was also used as an *eleothesium* for the rubbing and anointing process.

There are no modern buildings which can convey an idea of the extent and magnificence of these Roman

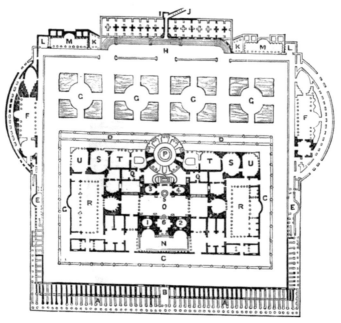

PLAN OF CARACALLA'S BATHS.

A Colonnade facing the street.
B Private bath rooms.
C Principal entrances.
D Internal corridors.
E Seats for bathers.
F Saloons for conversation.
G Open-air walks.
H *Theatridium*, or amphitheatre.
I Water reservoir.
J Aqueduct.
K, L, M Rooms for gymnastic exercises.

N Swimming bath.
O *Caldarium*, or hot-water bath.
P *Laconicum*, or vapour bath.
Q Supply cisterns.
R Covered halls.
S Cold-water bath.
T Room for sweet unctions.
U Cooling room.
1, 2, 3, 4 Private rooms.
5, 6 *Labra*, or public basins.

Thermæ, which were not only devoted to the purposes of bathing, but also comprised saloons for conversation or discussion, galleries of pictures and sculpture, libraries, walks planted with shady trees, porticos for gymnastic exercises, and, in fine, all that could contribute to the material and intellectual pleasure of a rich and luxurious people. The largest were those of Caracalla, situated near Mount Aventine, which measured about 675 yards in length, by 540 in width. They contained 1,600 seats of polished marble, and accommodation for no less than 2,300 bathers. On one side stood the temples of Apollo and Esculapius, the protectors of health ; and, on the other, those of Hercules and Bacchus, tutelary gods of the Antoninus family. They are still in a sufficient state of preservation to have enabled Pardini, the learned Italian architect, to make a plan of them, the foregoing copy of which may not prove uninteresting to my readers.

Although in all the baths there was a part set aside

for ladies, it was not so generally frequented as that used by the men, and the rich patrician matrons preferred attending to the duties of the toilet in their own houses. Indeed, this was no small matter for them, and with many it was the sole occupation ; hence the various implements appertaining to the toilet were styled *mundus muliebris*, or a woman's world.

The Woman.
(Ornatrix.)

Surrounded by a crowd of young slaves (called *cosmetæ*)

belonging to various nations, from the dark Nubian to the fair Gaul, who had each their particular department, and were marshalled by the *ornatrix*, or grand mistress of the toilet, the Roman lady sat in state, and made all tremble around her. Woe be to the unfortunate maid whose awkward fingers had not given a sufficiently graceful turn to her mistress's locks, or had not applied the paint to her cheek in its proper place. A pinch on the arm, a prick of a pin, or a heavy metal mirror hurled at her head, soon apprised her of the lady's displeasure. Juvenal, the bitter satirist of Roman manners, thus describes one of these scenes :—

Toilet Ewer.

> " She hurries all her handmaids to the task ;
> Her head alone will twenty dressers task ;
> Psecas, the chief, with neck and shoulders bare,
> Trembling, considers every sacred hair.
> If any straggler from his rank be found,
> A pinch must for the mortal sin compound.
> Psecas is not in fault ; but, in the glass,
> The dame's offended at her own ill face :
> The maid is banish'd, and another girl,
> More dext'rous, manages to comb and curl ;
> The rest are summon'd on a point so nice,
> And first the grave old woman gives advice ;
> The next is call'd, and so the turn goes round,
> As each for age or wisdom is renown'd.
> Such counsel, such deliberate care they take,
> As if her life and honour lay at stake ;

> With curls on curls they build her head before,
> And mount it with a formidable tow'r:
> A giantess she seems, but look behind,
> And then she dwindles to the pigmy kind."

There were three kinds of perfumes principally used by the Romans—the *hedysmata*, or solid unguents; the

Roman Comb.

stymmata, or liquid unguents, having an oily basis; and the *diapasmata*, or powdered perfumes. The unguents formed a numerous class, and their names were borrowed, some from the ingredients which entered into their composition, some from the original place of their production, and others, again, from the peculiar circumstances under which they were first made. Like our present preparations, they succeeded each other in public favour, and novelty was as great an attraction to the Roman belles as it is to our own modern ladies. There were the simple unguents, flavoured with one aroma, such as the *rhodium*, made from roses; the *melinum*, from quince blossoms; the *metopium*, from bitter almonds;

Roman Mirrors.

the *narcissinum*, from narcissus flowers; the *malobathrum*, prepared from a tree called so by Pliny, and supposed by some to be the *laurus cassia;* and many others too numerous to mention. The compound unguents were prepared by combining several ingredients. The most celebrated were the *susinum*, a fluid unguent, made of lilies, oil of ben, calamus, honey, cinnamon, saffron, and myrrh; the *nardinum*, made of oil of ben, sweet rush, costus, spikenard, amomum, myrrh, and balm; and, above all, Pliny praises the regal unguent, which was originally prepared for the king of the Parthians, and which consisted of no less than twenty-seven ingredients.[1] Some of these preparations were very costly, and sold for as much as four hundred denarii per pound, or about £14. The Romans not only applied them to the hair, but to the whole of the body, even to the soles of their feet. The most refined, indeed, adopted, as did the Grecian epicures, a different perfume for each part of their person. Besides this, their baths, their cloths, their beds, the walls of their houses, and even their military flags, were impregnated with sweet odours. Some carried this taste so far as to rub their horses and dogs with scented ointment.

Saffron was one of the perfumes most in favour with the Romans. They not only had their apartments and banqueting-halls strewed with this plant, but they also composed with it unguents and essences which were highly prized. Some of the latter were often made to flow in small streams at their entertainments, or to descend in odorous dews over the public from the *velarium* forming the roof of the amphitheatre. Lucan, in his "Pharsalia,"[2] describing how the blood runs out of the veins of a person bitten by a serpent, says that it spouts out in the same manner as the sweet-smelling essence of saffron issues from the limbs of a statue.

[1] Pliny's Nat. His., b. xiii. chap. 2.   [2] Lucan, Pharsal., b. ix. v. 809.

Perfumes were usually inclosed in bottles (*unguent-aria*) made of alabaster, onyx, or glass, of the shapes copied below from specimens in the Naples Museum. When required for the bath, they were carried in a round ivory box, called *narthecium*, like this engraving, copied from one found at Pompeii. Common perfumes were sold in little gilt shells,[1] or vessels, made of clay.

Scent-box.
(Narthecium)

The Roman perfumers (called *unguentarii*) were very numerous, and occupied a part of the town named *vicus thuraricus* in the Velabrum. The most celebrated in Martial's time was Cosmus, whom he frequently mentions in his Epigrams.[2] In Capua, a city

Roman Perfume Bottles (Unguentaria).

noted for its luxury, the perfume vendors occupied a whole street of the town, called Seplasia. They extracted some of their essences from flowers grown in Italy, but most of their ingredients were imported from Egypt and Arabia; and some of them were so costly, that the slaves who worked in their laboratories were stripped before they went home, to see that they had none concealed about them.

[1] Martial, b. 3, lxxxii   [2] Ibid., b. 1, lxxxvii.; b. 3, lv.

The custom of using perfumes in the *triclinium*, or dining-room, had been transmitted by the Greeks to the Romans; and the latter carried it, perhaps, to a still greater extent, for no banquet was considered complete without them, and they formed an indispensable item in the "bill of fare." Catullus, when inviting Fabullus to supper, after enumerating the various treats he has in store for him, adds—

Triclinium found at Pompeii.

> " And I can give thee essence rare
>   That Loves and Graces gave my fair;
>   So sweet its odour flows,
>   Thou'lt pray the gods ' May touch and taste
>   Be quite in smell alone effaced,
>   And I become all nose.' "

Martial does not appear to have enjoyed the happy state preconised by Catullus, of being "all nose;" for, in one of his epigrams, he complains to his host for giving him more perfumes than viands, thus reducing him to the state of a living mummy :—

> " Faith ! your essence was excelling,
>   But you gave us nought to eat;
> Nothing tasting, sweetly smelling,
>   Is, Fabullus, scarce a treat.

> " Let me see a fowl unjointed
>   When your table next is spread;
> Who not feeds, but is anointed,
>   Lives like nothing but the dead."

The witty critic was evidently not one of Cosmus's best customers, for he often ridicules the use of perfumes, saying that

> "He that smells always well does never so." [1]

And addressing Polla, an old coquette who sought by artifice to conceal the ravages of time, he exclaims—

> "Leave off thy paint, perfumes, and youthful dress,
> And nature's failing honestly confess.
> Double we see those faults which art would mend;
> Plain downright ugliness would less offend." [2]

The following picture of a "*beau*" of the period shows that ladies were not alone addicted to an extravagant use of perfumes :—

> "A beau is one who with the nicest care
> In parted locks divides his curling hair;
> One who with balm and cinnamon smells sweet,
> Whose humming-lips some Spanish air repeat;
> Whose naked arms are smoothed with pumice-stone,
> And toss'd about with graces all his own." [3]

In addition to the liquid essences and unguents, the Romans made use of an immense variety of cosmetics for improving and preserving the complexion. Pliny, in his " Natural History," gives a description of these preparations, some of which consisted of pea-flour, barley-meal, eggs, wine-lees, hartshorn, bulbs of narcissus, and honey; others simply of corn-flour, or crumb of bread soaked in milk. They made with these pastes a sort of poultice, which they kept on the face all night and part of the day. Some, indeed, only removed them for the purpose of going out, and Juvenal tells us, in one of his satires, that a Roman husband of his time

[1] Martial, b. 1, xii.   [2] Ibid., b. 3, xlii.   [3] Ibid., b. 3, lxiii.

seldom sees his wife's face at home, but when she sallies forth—

> "Th' eclipse then vanishes ; and all her face
> Is open'd and restored to every grace ;
> The crust removed, her cheeks as smooth as silk
> Are polish'd with a wash of asses' milk ;
> And should she to the farthest North be sent,
> A train of these attend her banishment." [1]

The last lines allude to Poppæa, the wife of Nero, who used to bathe in asses' milk every day, and when she was exiled from Rome, obtained permission to take with her fifty asses to enable her to continue her favourite ablutions.

Ovid, the poet of love, wrote a book on cosmetics,[2] of which, unfortunately, but a fragment came down to us. I shall give one or two extracts from it, if only to afford ladies who may be curious in these matters an opportunity of testing the virtues of the recipes given by the poet.

"Learn from me the art of imparting to your complexion a dazzling whiteness, when your delicate limbs shake off the trammels of sleep. Divest from its husk the barley brought by our vessels from the Libyan fields. Take two pounds of this barley with an equal quantity of bean-flour, and mix them with ten eggs. When these ingredients have been dried in the air, have them ground, and add the sixth part of a pound of hartshorn, of that which falls in the spring. When the whole has been reduced to a fine flour, pass it through a sieve, and complete the preparations with twelve narcissus bulbs pounded in a mortar, two ounces

[1] Juvenal, Sat. vi.   [2] Medicamina Faciei.

of gum, as much of Tuscan seed, and eighteen ounces of honey. Every woman who spreads this paste on her face will render it smoother and more brilliant than her mirror."

Another recipe he gives for removing blotches from the complexion consists in a mixture of roasted lupines, beans, white lead, red nitre, and orris-root, made into a paste with Attic honey.

Frankincense he also recommends as an excellent cosmetic, saying that if it is agreeable to gods, it is no less useful to mortals. Mixed with nitre, fennel, myrrh, rose-leaves, and sal ammoniac, he gives it as an excellent preparation for toilet purposes.

Besides these, the Romans also used *psilotrum*, a sort of depilatory, white lead or chalk for the face, *fucus*, a kind of rouge for the cheeks, Egyptian kohl for the eyes, barley-flour kneaded with fresh butter to cure pimples, calcined pumice-stone to whiten the teeth, and various sorts of hair dyes. Of the latter, the most curious

was a liquid for turning the hair black, prepared from leeches which had been left to putrefy during sixty days in an earthen vessel with wine and vinegar. As, however, blondes were very scarce among the Roman ladies, the most fashionable dye was

Roman Lady applying Fucus. one which changed their naturally dark hair to a sandy or fair colour. This was principally accomplished by means of a soap from

Gaul or Germany, called *sapo* (from the old German *sepe*), and composed of goat's fat and ashes. It is rather remarkable that this was the first introduction of soap we find mentioned, and that it was then solely applied to the purpose of dyeing the hair. Martial designates this dye under the name of Mattiac balls,[1] because they came from Mattium, a town of Germany, supposed to be Marpurg, and sarcastically sends them to an octogenarian, who is completely bald, to change the colour of his hair.

There is no doubt that some of these preparations were very injurious to the hair; for Ovid, in one of his elegies,[2] reproaches his mistress with having destroyed her flowing locks by means of dyes. "Did I not tell you to leave off dyeing your hair? Now you have no hair left to dye. And yet nothing was handsomer than your locks. They came down to your knees, and were so fine that you were afraid to comb them." Then he adds, a little further, "Your own hand has been the cause of the loss you deplore; you poured the poison on your own head. Now Germany will send you slave's hair; a vanquished nation will supply your ornament. How many times, when you hear people praising the beauty of your hair, you will blush and say to yourself, 'It is a bought ornament to which I owe my beauty, and I know not what Sicamber virgin they are admiring in me! And yet there was a time when I deserved all these compliments.'"

In such cases, as will be seen from the preceding

[1] Martial, b. 14, xxvii. [2] Elegy xiv.

extract, false hair was resorted to ; but baldness was not always the excuse for wearing such an appendage. The rage for *blonde* hair was so great at one time, that when ladies did not succeed in imparting the desired shade to their naturally raven tresses, they cut them off, to replace them with flaxen wigs. This was probably what had been done by the lady referred to by Martial :—

> "The golden hair that Galla wears
>   Is hers: who would have thought it?
> She swears 'tis hers, and true she swears,
>   For I know where she bought it."

That false hair was in fashion with ladies may be judged from the fact that even busts like that of Julia Semi-

amira, mother of Heliogabalus, represented here, were made with wigs of a different coloured marble, which could be removed at pleasure.

Ladies were not, however, the only

Julia Semiamira.   ones who tampered with their locks. The sterner sex did not disdain to practise this deceit ; and Martial, apostrophizing one of these chameleons in human garb, asks him how it is that he who was a " swan before, has now become a crow."

The Roman matrons were not less expert and tasteful than the Greek ladies in their modes of dressing the hair ; but their coiffures, like their perfumes, were principally borrowed from the latter. Thus we find the Grecian *strophos* adopted by the Romans under the name of *vitta*. This pretty head-dress, which has been lately revived amongst us, consisted of simple bands wound round the hair. It was confined to young maids, and was strictly forbidden to persons of bad character, who usually wore the *mitra*

Roman Head-dresses.

Tutulus. Nimbus. Vitta.

mentioned in the last chapter. The net was again patronised under the name of *reticulum;* and the only two head-dresses of strictly Roman creation were perhaps the *tutulus* and the *nimbus*, both of which are represented here. Some simply wore a long pin (*acus*), to hold the hair at the back of the head.

Hair-pin. (Acus).

When a man attained his majority and assumed the toga, he shaved his beard and offered it to some god. Nero presented his in a golden box set with pearls to Jupiter Capitolinus. Shaving continued in fashion until the time of Emperor Hadrian, who, to cover some excrescences on his chin, revived the custom of letting the beard grow, which his courtiers naturally hastened to adopt. How many modern fashions can thus be traced to the caprice or convenience of some influential person !

False hair was worn by men as well as by women;

and if we are to credit Suetonius, the Roman *perruquiers* had attained some proficiency in the art; for he tells us that Otho's wig was so cleverly made that it looked perfectly natural. These appendages, however, were very costly at that time, and a certain Phœbus, who had probably more imagination than ready cash, and could not afford to treat himself to an "invisible peruke," had drawn on his bald pate imaginary locks by means of a dark pomatum, whereupon Martial thus apostrophises him in his usual sarcastic style :—

> " Phœbus belies with oil his absent hairs,
> And o'er his scalp a painted peruke wears;
> Thou need'st no barber to dress thy pate,
> Phœbus ; a sponge would better do the feat." [1]

[1] Martial, b. 6, lvii.

A PERFUME BAZAAR IN THE EAST.

CHAPTER VII.

The Orientals.

Know ye the land of the cedar and vine,
Where the flowers ever blossom, the beams ever shine;
Where the light wings of Zephyr, oppress'd with perfume,
Wax faint o'er the gardens of Gúl in her bloom!
Where the citron and olive are fairest of fruit,
And the voice of the nightingale never is mute.
. . . .
'Tis the clime of the East; 'tis the land of the Sun.   Byron.

L.BOURDELIN

UXURIES are only sought and enjoyed by people living in a high state of refinement. When the Roman Empire of the West crumbled beneath the attacks of a horde of barbarians, who invaded its fertile plains and laid waste its magnificent cities, the arts of civilization, which they were unable to appreciate, took refuge in the Eastern metropolis where they had been cultivated since the days of Constantine the Great. Among these arts perfumery was

ranked, and the Greek emperors and their court showed
for aromatics a fondness at least equal to that which
had been displayed by their Western predecessors.
Having at their command all the fragrant treasures
of the East, they made a lavish use of them in private
life, and in all public festivals perfumes were made to
play an important part. Nor were they confined to
profane purposes, for the Oriental Church had likewise
introduced them into all their religious ceremonies, and
their consumption was so large at one time that the
priests purchased in Syria a piece of ground ten square
miles in extent, and planted it with frankincense-trees
for their own special requirements.

After several centuries of glory and splendour, the
Eastern Empire, torn by religious dissensions, was
doomed in its turn to fall under the aggressions of its
enemies, and although it struggled many years against
the followers of Mahomet, the Crescent succeeded at
last in replacing the Cross on the proud domes of Constantinople. In this instance, however, the conquerors
were nearly as polished as the vanquished. If their
religion, by forbidding them to delineate the form of
man in any way, had checked their progress in art, it
offered no impediment to the pursuit of science, and
they had already attained considerable proficiency in
many of its most important branches. To the Arabs,
indeed, we are indebted for many valuable discoveries
in the field of knowledge, and these children of the
desert may well be called the connecting link between
ancient and modern civilisation.

Avicenna, an Arabian doctor who flourished in the tenth century, was the first to study and apply the principles of chemistry, which was but imperfectly known to the ancients. This extraordinary man, who in a wandering life of fifty-eight years found time to write nearly one hundred volumes (twenty of which were a General Encyclopædia), is said to have invented the art of extracting the aromatic or medicinal principles of plants and flowers by means of distillation.[1] Perfumes had for many years been known and used by his countrymen, and long before Mahomet's time, Musa, one of the chief cities in Arabia Felix, was a celebrated emporium for frankincense, myrrh, and other aromatic gums; but hitherto the far-famed "perfumes of Araby the blest" had merely consisted in scented resins and spices. The floral world, so rich and fragrant in those favoured climes, had not yet been made to yield its sweet but evanescent treasures. To Avicenna belongs the merit of saving their volatile aroma from destruction and rendering it permanent by means of distillation.

The Orientals always exhibited for the rose a partiality almost equal to that of the nightingale, who is said to dwell constantly among its sweet bowers. It was, therefore, on that flower that Avicenna made his first experiments, selecting the most fragrant of the species, the *Rosa centifolia*, called by the Arabs, *Gul sad berk*.

---

[1] The word *al-embic*, which was formerly used in England and is still used in France to designate a *still*, clearly shows its Arabian origin.

> " The floweret of a hundred leaves,
>   Expanding while the dew-fall flows,
>   And every leaf its balm receives." [1]

He succeeded by his skilful operations in producing
the delicious liquid known as rose-water, the formula
for which is to be found in his works and in those of
the succeeding Arabian writers on chemistry.   It soon
came into general use, and appears to have been manufactured in large quantities, if we are to believe the
historians, who tell us that when Saladin entered Jerusalem in 1187, he had the floor and walls of Omar's
mosque entirely washed with it.

Rose-water is still held in high repute in the East,
and when a stranger enters a house the most grateful
token of welcome which can be offered to him is to
sprinkle him over with rose-water, which is done by
means of a vessel with a narrow spout, called *gulabdan*.
It is to this custom that Byron alludes in " The Bride
of Abydos," when he says—

> " She snatched the urn wherein was mix'd
>   The Persian Atar-gul's perfume,
>   And sprinkled all its odours o'er
>   The pictured roof and marbled floor.
>   The drops that through his glitt'ring vest
>   The playful girl's appeal address'd,
>   Unheeded o'er his bosom flew,
>   As if that breast was marble too."

Niebuhr, in his " Description of Arabia," mentions
likewise this habit of throwing rose-water on visitors as
a mark of honour, and says it is somewhat amusing to
witness the discomfited and even angry looks with
which foreigners are wont to receive these unexpected

[1] Moore's Lalla Rookh.

aspersions. The censer is also generally brought in afterwards, and its fragrant smoke directed towards the beards and garments of the visitors, this ceremony being considered as a gentle hint that it is time to bring the visit to an end.[1]

According to the same authority, Arabian censers are made of wood (probably lined with metal) and covered with plaited cane, like the specimen represented here.

Arabian Censer and Gulabdan.

The *gulabdan*, or "casting bottle," as it was called in this country two or three centuries back, is either of glass or earthenware in ordinary houses, but among rich people both these implements are of gold or silver richly chased or ornamented. The engraving on next page illustrates this important feature in Oriental customs. The female servant carrying the perfume-burner and sprinkling-vase is taken from La Mottraye's print of a Turkish harem, and the man from a picture in the late Lord Baltimore's collection representing the reception of a French ambassador by the Grand Vizier. The per-

[1] Niebuhr, Description de l'Arabie.

fumes used in the censer combine all the fragrant
woods and gums of the East, among which the aloe,
mentioned in Chapter III., stands prominent:

> " The aloes-wood, from which no fragrance came,
>   If placed on fire, its inodorous state
>   Will change, more sweet than ambergris." [1]

Mahomet, who was a keen observer of human nature,
founded his religion on the enjoyment of all material

Turkish Servants bearing Perfumes.

pleasures, well knowing that it was the best means of
securing the adhesion of his sensual countrymen. He had
forbidden, it is true, the use of wine, but simply because he
feared the dangerous excesses to which it gave rise: the
indulgence in perfumes was one, on the contrary, he liked
to encourage, for they assisted in producing in his adepts
a state of religious ecstasy favourable to his cause.   He

[1] Sâdī's Gulistān, chap. i. st. 18.

professed himself a great fondness for them, saying that what his heart enjoyed most in this world were children, women, and perfumes, and among the many delights promised to the true believers in the *Djennet Firdous,* or Garden of Paradise, perfumes formed a conspicuous part, as will be seen from the following description, taken from the Koran :—

When the day of judgment comes, all men will have to cross a bridge called Al Sirat, which is finer than a hair, and sharper than the edge of a Damascus blade. This bridge is laid over the infernal regions, and however dangerous and difficult this transit may appear, the righteous, upheld and guided by the prophet, will easily accomplish it; but the wicked, deprived of such assistance, will slip and fall into the abyss below, which is gaping to receive them.

After having passed this first stage, the "right-hand men," as the Koran calls them, will refresh themselves by drinking at the pond of Al Cawthar, the waters of which are whiter than milk or silver, and more odoriferous than musk. They will find there as many drinking-cups as there are stars in the firmament, and their thirst will be quenched for ever.

They at last will penetrate into Paradise, which is situated in the seventh heaven, under the throne of God. The ground of this enchanting place is composed of pure wheaten flour mixed with musk and saffron; its stones are pearls and hyacinths, and its palaces built of gold and silver. In the centre stands the marvellous tree called *tuba,* which is so large that a man mounted on the fleetest horse could not ride round its branches in one hundred years. This tree not only affords the most grateful shade over the whole extent of Paradise, but its boughs are loaded with delicious fruit of a size and taste unknown to mortals, and bend themselves at the wish of the inhabitants of this happy abode.

As an abundance of water is one of the greatest desiderata in the East, the Koran often speaks of the rivers of Paradise as one of its chief ornaments. All those rivers take their rise from the tree *tuba :* some flow with water, some with milk, some with honey, and others even with wine, this liquor not being forbidden to the blessed.

Of all the attractions, however, of these realms of bliss, none will equal their fair inhabitants—the black-eyed houris[1]—who will welcome the brave to their bowers, waving perfumed scarves before them,[2] and repaying with smiles and blandishments all their toils and fatigues. These beauteous nymphs will be perfection itself in every sense: they will not be created of our own mortal clay, but of *pure musk.*

I doubt very much if the prospect of inhabiting a place with a soil of *musk,* peopled with ladies composed of the same material, would prove a great allurement to our Europeans, with their nervous tendencies ; the bare notion of such a possibility would be sufficient to give a

---

[1] "Houri" comes from the words *hur al oyoun,* "the black-eyed."

[2] "Waving embroider'd scarves whose motion gave
    Perfume forth, like those the Houris wave
    When beckoning to their bowers the Immortal Brave."
                                    MOORE'S *Lalla Rookh.*

headache to some of the more sensitive. But in the East tastes are different; and it is a singular fact that the warmer a country is, the greater is the taste for strong perfumes, although one would suppose that the heat, developing to the utmost such powerful aromas, would render them actually unbearable.

As an instance of the fondness which the Orientals exhibit for musk, Evlia Effendi relates that in Kara Amed, the capital of Diarbekr, there is a mosque called *Iparie*, built by a merchant, and so called because there were mixed with the mortar used in its construction seventy juks of musk, which constantly perfume the temple. The same author describes the mosque of Zobaide, at Tauris, as being constructed in a similar way: and as musk is the most durable of all perfumes, the walls still continue giving out the most powerful scent, especially when the rays of the sun strike upon them.

Many of Mahomet's prescriptions were of a sanitary nature, and in order to insure their observance by his superstitious followers, he gave them, like Moses, the form of religious laws. Such were the ablutions and purifications ordained by the Koran.[1] All true believers are strictly enjoined to wash their heads, their hands as far as the elbows, and their feet as far as the knees, before saying their prayers; and when water is not to be procured, fine sand is to be used as a substitute.

When the Turks settled themselves in the Greek Empire, they did not rest satisfied with these limited ablutions, but soon adopted the luxurious system of

[1] Koran, v. 8, 9.

baths which they found already established in the con-
quered cities. These baths have been fully described
in the last chapter; they have, moreover, been lately
introduced into London; and although what we are
offered is but a pale copy of the magnificence of the
palaces devoted to that purpose in the East, it might be

Turkish Bath.

thought superfluous to dwell any longer on this subject.
The above illustration will suffice to convey an idea of
the style of these buildings.

Soap is sometimes used in these establishments, but
they more frequently employ a sort of saponaceous clay
scented with the sweetest odours, which is, no doubt, a
lineal descendant of that *smégma* mentioned in the Greek
chapter as being in great favour among the Athenians.
It is to that preparation that Sâdī, the celebrated Per-

sian poet, alludes in the following beautiful apologue, whereby he illustrates the benefit of good society :—

> "'Twas in the bath, a piece of perfumed clay
> Came from my loved one's hand to mine, one day.
> 'Art thou, then, musk or ambergris ?' I said ;
> 'That by thy scent my soul is ravished ?'
> 'Not so,' it answered, 'worthless earth was I,
> But long I kept the rose's company ;
> Thus near, its perfect fragrance to me came,
> Else I'm but earth, the worthless and the same.'"[1]

The rose, as I said before, is the favourite flower of the Orientals. The beauty of its aspect and the sweetness of its perfume are favourite themes for their poets. The finest poem that ever was written in the Persian language, the "Gulistān," from which the above is extracted, means the garden of roses, and Sādī, its author, with the *naïve* conceit of Eastern writers, thus explains his motives for giving that name to his work :—

"On the first day of the month of Urdabihisht (May), I resolved with a friend to pass the night in my garden. The ground was enamelled with flowers, the sky was lighted with brilliant stars ; the nightingale sang its sweet melodies perched on the highest branches ; the dew-drops hung on the rose like tears on the cheek of an angry beauty ; the parterre was covered with hyacinths of a thousand hues, among which meandered a limpid stream. When morning came my friend gathered roses, basilisks, and hyacinths, and placed them in the folds of his garments ; but I said to him, 'Throw these

---

[1] Sādī's Gulistān, Pref.

away, for I am going to compose a Gulistān (garden of roses), which will last for eternity, whilst your flowers will live but a day.' "

Hafiz, another renowned Persian poet, was also a great admirer of flowers and perfumes, which are constantly recurring in his verses, and furnish him with the most charming similes. Addressing his mistress in one of his *Gazels*, he exclaims—

> " Like the bloom of the rose, when fresh pluck'd and full blown,
>   Sweetly soft is thy nature and air :
> Like the beautiful cypress in Paradise grown,
>   Thou art ev'ry way charming and fair.

> " When my mind dwells on thee, what a lustre assume
>   All the objects which fancy presents !
> On my memory thy locks leave a grateful perfume,
>   Far more fragrant than jasmine's sweet scents." [1]

Hafiz seems, like Anacreon, to have particularly worshipped the rose ; and, as his Grecian predecessor, he always couples in his odes the praise of wine with that of the queen of flowers :—

> " In the mirth-enliven'd bower,
>   Wine, convivial songsters, pour :
> See the garden's flowery guest
> Comes in happiness full dress'd ;
> Joy round us sweet perfume throws,
> Offspring of the blooming rose.

> " Hail ! sweet flower, thy blossom spread,
>   Here thy welcome fragrance shed ;
> Let us with our friends be gay,
> Mindful of thy transient stay :
> Pass the goblet round ; who knows
> When we lose the blooming rose ?

[1] Hafiz, Gazel xi.

" Hafiz loves, like Philomel,
   With the darling rose to dwell :
   Let his heart a grateful lay
   To her guardian [1] humbly pay,
   Let his life with homage close,
   To the guardian of the rose." [2]

That perfumes have been in use in the East, to please the living and honour the dead, since a very remote period, we find a proof in the following story, extracted from a Persian writer, relating the death of Yezdijird, the last of the Kaiānian race of kings, in the year 652.

That unfortunate monarch having fled from his dominions and taken refuge in the territory of Merv, its inhabitants were anxious to apprehend and destroy him ; they accordingly sent a message to Tanjtākh, king of Tartary, offering to place themselves under his protection, and to deliver the fugitive into his hands. Tanjtākh accepted their proposal and marched against Merv with a large army ; hearing which, Yezdijird left the cāravānserāi where he had alighted, and wandered about unattended in quest of a hiding-place. He at last came to a mill, where he begged for a night's shelter. The miller promised him that he should be unmolested; but his attendants having remarked that he was richly clad, murdered him in his sleep, and divided the spoil among themselves.

The next day Tanjtākh arrived at Merv, and caused Yezdijird to be sought in every direction. Some of the emissaries came to the mill, and having remarked that one of the servants smelt strongly of perfume, they tore

[1] The nightingale.   [2] Hafiz, Gazel ii.

open his garments, and found Yezdijird's imperial robe,
scented with otto of roses and other essences, hid in
his bosom. The body of the king was discovered in
the mill-dam, and brought before Tanjtākh, who wept
bitterly, and ordered it to be embalmed with spices and
perfumes, and buried with regal honours. The miller

A Persian Lady.

and his servants were put to death, in punishment for
their treachery.

The taste for perfumes has in no wise diminished
among modern Orientals; it has, on the contrary, been
constantly increasing, and now pervades all classes, who
seek to gratify it to their utmost, according to their

means. It is principally cultivated among ladies who, caring little or nothing for mental acquirements, and debarred from the pleasures of society, are driven to resort to such sensual enjoyments as their secluded mode of life will afford. They love to be in an atmosphere redolent with fragrant odours that keep them in a state of dreamy languor which is for them the nearest approach to happiness. The sole aim of their existence being to please their lords and masters, the duties of the toilet are their principal and favourite occupation. Many are the cosmetics brought into request to enhance their charms, and numerous are the slaves who lend their assistance to perform that important task, some correcting with a whitening paste the over-warm tint of the skin, some replacing with an artificial bloom the faded roses of the complexion.

> "While some bring leaves of henna, to imbue
> The fingers' ends with a bright roseate hue,
> So bright that in the mirror's depth they seem
> Like tips of coral branches in the stream;
> And others mix the kohol's jetty dye
> To give that long dark languish to the eye
> Which makes the maids whom kings are proud to cull
> From fair Circassia's vales so beautiful." [1]

Although, according to our European notions, redtipped fingers and darkened eyelids are not calculated to increase female loveliness, this may be looked upon as a mere conventional matter, and it may be fairly presumed that the constant cares which the Eastern ladies bestow on themselves have the effect of increas-

[1] Moore's Lalla Rookh.

ing and preserving their beauty. This is confirmed
by most travellers, and, among others, Sonnini in his
Travels in Egypt thus expresses himself on that
subject :—

"There is no part of the world where the women
pay a more rigid attention to cleanliness than in those
Oriental countries. The frequent use of the bath, of
perfumes, and of everything tending to soften and
beautify the skin and to preserve all their charms, em-
ploys their constant attention. Nothing, in short, is
neglected, and the most minute details succeed each
other with scrupulous exactness. So much care is not
thrown away; nowhere are the women more uniformly
beautiful, nowhere do they possess more the talent of
assisting nature, nowhere, in a word, are they better
skilled or more practised in the art of arresting or
repairing the ravages of time, an art which has
its principles and a great variety of practical recipes." [1]

As it may interest some of my fair readers to know
the composition of those far-farmed Oriental cosmetics,
I shall transcribe here the recipes of some of those preparations, for the authenticity of which I can vouch,
having received them from one of my correspondents at
Tunis,[2] to whom they were given by a native Arabian
perfumer. If not useful, they will no doubt be found
amusing.

The kohl, or kheul, which we have seen in use for

[1] Sonnini's Travels in Upper and Lower Egypt, p. 180.
[2] M. A. Chapelié.

darkening the eyelids since the time of the ancient Egyptians, is made by them in the following way :— They remove the inside of a lemon, fill it up with plumbago and burnt copper, and place it on the fire until it becomes carbonised; then they pound it in a mortar with coral, sandal-wood, pearls, ambergris, the wing of a bat, and part of the body of a chameleon, the whole having been previously burnt to a cinder and moistened with rose-water while hot.

A complexion-powder called *batikha*, which is used in all the harems for whitening the skin, is made in the following manner :—They pound in a mortar some cowrie-shells, borax, rice, white marble, crystal, tomata, lemons, eggs, and helbas (a bitter seed gathered in Egypt); mix them with the meal of beans, chick-peas, and lentils, and place the whole inside a melon, mixing with it its pulp and seeds; it is then exposed to the sun until its complete desiccation, and reduced to a fine powder.

The preparation of a dye used for the hair and beard is no less curious. It is composed of gall-nuts fried in oil and rolled in salt, to which are added cloves, burnt copper, minium, aromatic herbs, pomegranate flowers, gum-arabic, litharge, and henna. The whole of these ingredients are pulverised and diluted in the oil used for frying the nuts. This gives it a jet-black colour, but those who wish to impart a golden tint to their hair employ simply henna for that purpose.

That hair-dyes have been used in the East for many centuries appears from the following lines, in which

Sâdî ridicules the habit with a sarcastic spirit worthy of Martial :—

> "An aged dame had dyed her locks of grey ;
>   'Granted,' I said, 'thy hair with silver blent
> May cheat us now ; yet, little mother ! say,
>   Canst thou make straight thy back, which time has bent.' " [1]

To conclude the list of Oriental cosmetics, I may mention an almond paste, called *hemsia*, which is used as a substitute for soap ; a tooth-powder named *souek*, made from the bark of the walnut-tree ; pastilles of musk and amber paste (*kourss*), for burning and also for forming chaplets of beads, which the fair odalisques roll for hours in their hands, thus combining a religious duty with a pleasant pastime ; a depilatory called "termentina," which is nothing more than turpentine thickened into a paste ; and last, not least, the celebrated *schnouda*, a perfectly white cream, composed of jasmine pomade and benzoin, by means of which a very natural but transient bloom is imparted to the cheeks.

The far-famed Balm of Mecca is still greatly esteemed amongst the Orientals, and some even pretend that the limited quantity of the genuine article produced yearly is reserved for the Grand Seignior's special use. Lady Mary Wortley Montagu does not appear to have shared their admiration for it, for she relates in her letters that having had a small quantity presented to her, she applied it to her face, expecting some wonderful improvement from it, instead of which it made it red and swollen for three days.[2]

---

[1] Sâdî's Gulistân, chap. vi. st. 5.   [2] Lady Montagu's Letters, xxxvii.

The same authority furnishes us with a very accurate description of the Eastern mode of wearing the hair; and, as fashions are not so liable to change there as they are here, we may assume it as applicable to the present period. "The head-dress," says Lady Montague,[1] "is composed of a cap called *talpock*, which is, in winter, of fine velvet, embroidered with pearls or diamonds, and in summer of a light shining silver stuff. This is fixed on one side of the head, hanging a little way down with a gold tassel, and bound on either with a circle of diamonds or a rich embroidered handkerchief. On the other side of the head the hair is laid flat, and here the ladies are at liberty to show their fancies, some putting flowers, others a plume of heron's feathers, and, in short, what they please; but the most general fashion is a large bouquet of jewels made like natural flowers—that is, the buds of pearl, the roses of different coloured rubies, the jessamines of diamonds, the jonquils of topazes, etc., so well set and enamelled, 'tis hard to imagine anything of that kind so beautiful. The hair hangs at its full length behind, divided into tresses braided with pearl and ribbon, which is always in great quantity."

The Turks shave their heads, leaving a single tuft of hair on the top, by which they expect Azrael, the angel of death, to seize them when conveying them to their last abode. They preserve their beard with the greatest care, and make it a point of religion to let it grow, because Mahomet never cut off his. No greater insult can

[1] Lady Montague's Letters, xxix.

be offered to a Mahometan than to deprive him of this hirsute ornament ; it is a degradation reserved for slaves, or a punishment inflicted on criminals.

The barber of the King of Persia is no insignificant personage ; he enjoys all the privileges and considera- tion naturally attached to one who has in his charge such a venerated object as a royal beard. The *dellak*, or barber, of the great Schah Abbas amassed such riches that he built a splendid bridge, which still bears his name ; and his modern successor erected, not long since, a magnificent palace for himself in the vicinity of the Royal Baths at Teheran.

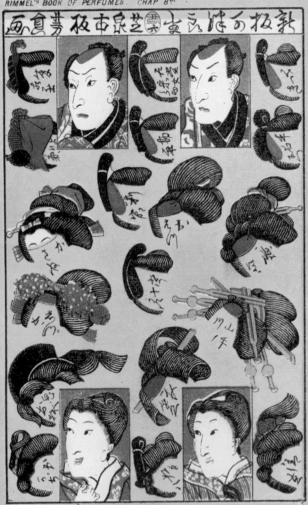

FAC-SIMILE OF A
JAPANESE FASHION PLATE

CHAPTER VIII.

THE FAR EAST.

"Be like the perfume-sellers, for thy dress
Near them will share the odours they possess."
PILPAY'S INDIAN APOLOGUES.

ONTINUING our peregrinations "all round the world," we now come to the Far East, that fairy-land of the ancients which we more sober - minded moderns simply designate under the names of India, China, and Japan. Here our history will cease to be chronological, for the arts of civilisation have been known and practised by those nations from a very remote period, and little if any would be the change or progress to be traced among them for many centuries.

To commence with India, we find that perfumes have been used in that country since the earliest records; a fact easily accounted for by the sensual temperament of its inhabitants, and the abundance of fragrant materials placed at their disposal by bountiful Nature. Kálidása, a Sanskrit writer, who flourished under the reign of king Vikramáditya I., some two thousand years ago, frequently mentions perfumes in his poems, and especially in the beautiful drama called "Śakoon-talá; or, the Lost Ring." From him we learn they were applied both to sacred and private purposes.

Sacrifices were usually offered in the temples of the Indian Trinity, or Tremoortee, comprising Brahmá, Vishņu, and Śiva. According to the Vedás they were to consist of a fire of fragrant woods lighted at each of the four cardinal points. The flames were fed now and then with a consecrated ointment, and around the fire was scattered a scented herb called *kúsa*,[1] which was held sacred. Kanwa, the father of Śakoontalá, who is the chief of the hermits, offers one of these sacrifices in the above-mentioned drama, and exclaims—

> "Holy flames that gleam around
> Every altar's hallowed ground;
> Holy flames, whose frequent food
> Is the consecrated wood,

[1] I believe this to have been the herb I found in the East India collection at the International Exhibition, under the name of *rusa*. It is the *Andropogon nardus*, or ginger grass (improperly called Indian geranium), from which an oil is extracted which is used in perfumery.

> And for whose encircling bed
> Sacred Kúsa-grass is spread ;
> Holy flames that waft to heaven
> Sweet oblations daily given,
> Mortal guilt to purge away ;
> Hear, oh, hear me, when I pray,
> Purify my child this day !" [1]

As will be seen from the last words of this prayer, sacrifices were not only offered by the Hindús as a general mode of worship, but also to propitiate the gods on particular occasions, as was done by the ancient Greeks and Romans. In this instance Śakoontalá is about to be married, and her father invokes the blessings of the deities upon her. These ceremonies did not always take place in temples, but sometimes in consecrated groves. In this same drama, King Dushyanta, alluding to this custom, says—

> "The sprouting verdure of the leaves is dimmed
> By dusky wreaths of upward-curling smoke
> From burnt oblations."

It was considered no sin to apply sacred grass to private purposes, for we find Anasúyá, one of Śakoontalá's handmaids, compounding perfumes and unguents with consecrated paste and this kúsa-grass, to anoint the limbs of her mistress, when attending to her bridal toilet.[2] Some of these preparations were believed to possess medicinal properties, and such was the ointment of Usira-root,[3] brought to the Indian beauty by another assistant as a cure for fever.

The custom of staining the soles of the feet with

---

[1] Śakoontalá, Act iv.   [2] Śakoontalá, Act iv.
[3] This root is probably the Indian kus-kus, or vetivert (*Anatherum muricatum*).

henna appears to have been very ancient, for we find
it mentioned by one of the hermits who brings bridal
presents for Śakoontalá, and thus describes a mysterious
forest where he found them :—

> " Straightway depending from a neighbouring tree
> Appeared a robe of linen tissue, pure
> And spotless as a moonbeam—mystic pledge
> Of bridal happiness ; another tree
> Distilled a roseate dye wherewith to stain
> The lady's feet."

In an Indian ode called " Megha-dúta," translated by
Paterson, there occurs, also, the following passage
alluding to the same fashion :—

> " The rose hath humbly bowed to meet
> With glowing lips her hallowed feet,
> And lent them all its bloom."

According to Hindú mythology there are five hea-
vens, over each of which presides one of their superior
gods. That of Brahmá, called Brahma-loka, is situated
on Mount Meru ; those of Vishṇu, Śiva, Kuvera, and
Indra are on the summit of the Himalayas. In all these
elysiums perfumes and flowers are among the chief delights. The principal ornament of Brahmá's heaven is

> " That blue flower which, Brahmins say,
> Blooms nowhere but in Paradise."

It is the blue campac or champac flower, a great rarity,
as the only sort known on this earth[1] has yellow blossoms with which Hindú girls are wont to ornament
their raven hair.

In Indra's paradise, called Swarga, is to be found the
still more attractive cámalatá, whose rosy flowers not

[1] *Michelia champaca.*

only enchant the senses of all those who have the happiness of breathing its delicious fragrance, but have also the power of granting them all they may desire. This Indra, the *Jupiter Tonans* of the Hindús, appears very partial to scent, for he is always represented with his breast tinged with sandal-wood.

Káma, the god of love, or Indian Cupid, is armed with a bow made of sugar-cane, the string of which consists of bees. He has five arrows, each tipped with the blossom of a flower, which pierce the heart through the five senses, and his favourite dart is pointed with the chú-ta or mango-flower. I regret to add that young maidens, with cruel dispositions,

Káma, the Indian Cupid.

hardly to be expected in their tender years, do not scruple to furnish the malicious god with weapons, as may be seen from the following quotation. A young maid plucks a mango-blossom and exclaims—

> " God of the bow, who with spring's choicest flowers
>   Dost point thy five unerring shafts ; to thee
>   I dedicate this blossom ; let it serve
>   To barb thy truest arrow ; be its mark
>   Some youthful heart that pines to be beloved."

A sweet little flower, mounted on a reed, does not appear at first sight to form a very dangerous weapon, yet it seems to inflict great pain, if we are to credit the complaints exhaled by a wounded swain, who says, in the same poem—

> " Every flower-tipped shaft
> Of Káma, as it probes our throbbing hearts,
> Seems to be barbed with hardest adamant."

Flowers and perfumes are still used in modern Hindú worship. Incense is burned in all ceremonies, and the temples are adorned with a profusion of fresh-gathered blossoms. Coloured ointments are also used to make hieratic signs on the face, arms, and chest. The sectaries of Vishṇu have a red and yellow line drawn horizontally on the forehead; those of Śiva wear the same line vertically. I saw in the East Indian collection at the last Exhibition some specimens of these ointments, which were very strongly flavoured with sandal-wood, and other indigenous essences. In a religious *fête* called *Mariatta Codam*, the devotees rub themselves over with an ointment made of saffron, and go round collecting alms, in return for which they distribute scented sticks, partly composed of sandal-wood, which are received with great veneration. At another held in honour of the goddess *Debrodee*, fakeers crowned with flowers sprinkle incense on glowing coals, which they place in their hands without appearing to experience any pain from it. At the Krishna festival a red powder diluted in rose-water is liberally distributed by means of syringes over all passers-by, to the utter discomfiture of their wearing apparel. A some-what similar custom is observed in the Birman Empire. On the 12th of April, which is the last day of their calendar, women throw water at all they meet, to wash away all the impurities of the past year and commence the new one free from sin. Rich people use rose-water mixed with sandal-wood for that purpose.

In Tibet incense is also burned, sometimes in a censer but more frequently in a gigantic altar, with an aperture at the top, which is called *Song-boom*, and bears some

Song-boom, or Tibetan Incense Altar.

resemblance to a lime-kiln.[1]   As, however, the fragrant gums of India are scarce in these northern regions, juniper is used as a substitute. They also make use in their worship of a very singular implement consisting

[1] Dr. Hooker's Himalayan Journal, vol. i. p. 339.

of a leather cylinder, which contains written prayers, and is turned with a handle. Each revolution causes a little bell to ring, and this counts for one prayer. Some people even think this mechanical mode of praying too fatiguing, and have their cylinders turned, like mills, by *water-power*.[1]

In Cochin China, when fishermen are about to start on a cruise, they seek to propitiate the deities of the perfidious element by burning aromatic and consecrated woods on altars formed of rude stones. The Javanese, who are the usual purveyors of those delicate birds' nests so highly prized by Chinese epicures, offer up likewise a sacrifice before venturing on these dangerous expeditions. They slaughter a buffalo, pronounce some prayers, anoint themselves with sweet-scented oils, and smoke with gum benzoin the entrance of the caverns where they are to seek the coveted prize. Near some of these caves a tutelar goddess is worshipped, whose priest burns incense, and lays his protecting hands on every person prepared to descend into the abyss.[2]

Hindú marriages are celebrated under a sort of canopy called *pendal*, which, among wealthy people, is richly ornamented and brilliantly lighted with lamps. The bride and bridegroom sit, or rather squat, at one end, and at the other burns the sacred fire or *oman*, which is constantly kept up by throwing into it sandal-wood, incense, scented oils, and other ingredients, which shed aromatic fumes. The Bráhmans, after having recited

[1] Dr. Hooker's Himalayan Journal, i. 195.
[2] Lord Macartney's Embassy to China.

a variety of prayers, consecrate the union of the couple by throwing a handful of saffron mixed with rice flour on their shoulders, and the ceremony ends by the husband presenting his wife with a little golden image called *talee*, which is worn round the neck by married women, as a substitute for the wedding-ring.[1]

Hindú Marriage Ceremony.

Scented woods are also used in the funeral piles which consume the remains of the dead, when the wealth of the deceased, or the generosity of his heirs, admits of such expense. When *suttees* were still in fashion, disconsolate widows could have the satisfaction of dying, like Sardanapalus, "stifled in aromatic smoke;" but

[1] L'Indoustan, vol. iii. p. 11.

since the British Government has abolished this custom
they are left to end their days like ordinary mortals.

There are few countries in the world equal to India
for the abundance and variety of its floral productions.

> " A hundred flowers there are beaming,
>   The verdure smiling and the hushed waves dreaming.
>   Each flower is still a brighter hue assuming,
>   Each a far league the love-sick air perfuming.
>   The rose her book of hundred leaves unfolding,
>   The tulip's hand a cup of red wine holding.
>   The northern zephyr ambergris round spreading,
>   Still through its limits varied scents is shedding." [1]

Whilst the southern provinces are rich with the
vegetation of tropical climes, the northern parts, and
especially Cashmere, teem with roses and other European flowers.

> " Who has not heard of the vale of Cashmere,
>   With its roses the brightest that earth ever gave,
>   Its temples and grottoes, and fountains as clear
>   As the love-lighted eyes that hang over their wave?" [2]

Otto of roses has been made for a very long time in
India, and Lieutenant-Colonel Polier thus relates its
origin in the " Asiatic Researches : "—" Noorjeehan
Begum (Light of the World), the favourite wife of
Jehan-Geer, was once walking in her garden, through
which ran a canal of rose-water, when she remarked
some oily particles floating on the surface. These were
collected, and their aroma found to be so delicious, that
means were devised to produce the precious essence in
a regular way."

Next in favour is the jasmine, which Hindú poets call

---

[1] Anvár-i Suhailí, ch. i. st. 26.   [2] Moore's Lalla Rookh.

the "Moonlight of the Grove." There are two species cultivated for their perfume—the *Jasminum grandiflorum*, or *Tore*, and the *Jasminum hirsutum*, or *Sambac*.

Among other fragrant flowers we may mention the Pandang (*Pandanus odoratissimus*), the Champac (*Michelia champaca*), the Kurna (*Phœnix dactilifera*), the Bookool (*Minusops elengi*), and last, not least, the Henna (*Lawsonia inermis*), the blossoms of which have a delicious odour.

From all these flowers essences are distilled, and the centre of this manufacture is Ghazepore, a town situated on the north bank of the Ganges above Benares. The process is extremely simple. The petals are placed in clay stills with twice their weight of water, and the produce is exposed to the fresh air for a night in open vessels. The next morning the otto is found congealed on the surface and is carefully skimmed off. These essences would be very beautiful if they were pure, but the native distillers being but little skilled in their art, add sandal-wood shavings to the flowers to facilitate the extraction of the otto, which thus becomes tainted with a heavy sandal-wood flavour. Besides these essences, perfumed oils are also made with some of these flowers in the following way:—Gingelly oil seeds are placed in alternate layers with fresh flowers in a covered vessel. The latter are renewed several times, after which the seeds are pressed, and the oil produced is found to have acquired the smell of the flowers. Musk, civet, ambergris, spikenard (*Valeriana Jatamansi*),[1] patchouly, and

[1] See chap. iii.

kus-kus are also favourite perfumes with the Indians. The last mentioned, which is the rhizome of the *anatherum muricatum*, is made into mats and blinds, which, being watered in the sun, give out a most pleasant odour.

Perfumes and flowers play a great part in Indian poetry, and the following extracts taken at random from "Anvár-i Suhailí"[1] will show to what happy comparisons they are applied :—

> " Like musk is moral worth ; from sight concealed
> 'Tis by its odour to the sense revealed."

" The damsel entered the king's chamber with a face like a fresh rose-bud which the morning-breeze has caused to blow, and with ringlets like the twisting hyacinth buried in an envelope of the purest musk."

> " With hyacinth and jessamine her perfumed hair was bound,
>   A posy of sweet violets her clustering ringlets seemed ;
> Her eyes with love intoxicate, in witching sleep half drowned,
>   Her locks to Indian spikenard like, with love's enchantments beamed."

The following description of a young maiden struck down by illness is exquisitely beautiful :—

" All of a sudden the blighting glance of unpropitious fortune having fallen on that rose-cheeked cypress, she laid her head on the pillow of sickness ; and in the flower-garden of her beauty, in place of the damask-rose, sprang up the branch of the saffron. Her fresh jessamine, from the violence of the burning illness, lost its moisture ; and her hyacinth full of curls, lost all its endurance from the fever that consumed her."

[1] Anvár-i Suhailí, or the Lights of Canopus, translated by E. B. Eastwick.

"Her graceful form, with lengthened sufferings spent,
Was like her perfumed musky tresses—bent."

The Hindú perfumer (called gund'hee) does not indulge, like his European *confrères*, in showy glass cases
and brilliant shops. His whole *establishment* consists
in a few sacks, boxes, and trays, containing his various
fragrant stores, in the midst of which he sits dispensing
them to his beauty-seeking patrons.

Gund'hee, or Hindú Perfumer.
(*From an original sketch.*)

The Hindú barber plies also his vocation in the
open air, and handles with great dexterity his razor,
mounted on hinges, which is a somewhat formidable
looking instrument. The specimen represented on the
next page is from an original in Mr. Berthoud's collection, which is of gilt metal, chiselled, and studded
with jewels.

My remarks have been hitherto confined to the Hindús, and although some of them will equally apply to the Mussulmans inhabiting India, the latter offer some peculiar characteristics which may be briefly described. In "Qanoon-e-Islam," a book written by Jaffur Shurreef, a native of the Deccan, is to be found some reliable information on this subject.

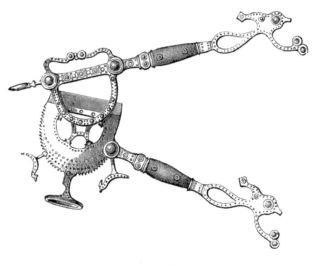

Hindú Razor.

Their customs naturally offer some resemblance to those of their Arabian ancestors, and their fondness for perfumes seems to have in no way decreased since the time of the Prophet. In all their ceremonies they burn عود *ood*, an incense composed of benzoin, aloe, sandalwood, patchouly, etc., and the *oodsoz*, or censer, is also lighted at the feet of the dead as soon as their eyes have been closed. صندل *sundul* or sandal-wood ointment is likewise used for religious purposes in so many instances that it would fill a book to relate them all. I shall merely quote one as being, perhaps, the most curious, and that

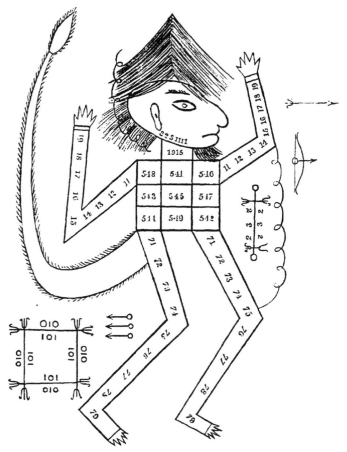

Magic Figure for Dawut or Exorcism.

is the *dawut* or exorcism. Magic circles, squares, and figures are drawn on a plank with *sundul*, and the individual supposed to be possessed with a demon is made to sit in the centre. The exorciser then pronounces an incantation in Arabic, and burns some incense under the nose of the patient, who is requested to inhale its fumes. It seems that demons are not partial to scents,

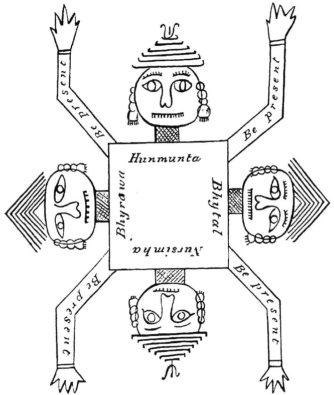

Magic Figure for Dawut or Exorcism.

for they generally allow themselves to be smoked out in this manner. The accompanying illustrations, which are perfectly authentic, represent two of these magic figures, which are supposed to bear some faint resemblance to the Evil One.

As an example of the lavish use of perfumes they make in private life, I may give a description of the *Singardan*, or toilet-bag, forming part of the presents which a bridegroom usually sends to his bride elect. This *nécessaire* contains, among other things, a *pandan*, or box to hold betel, an aromatic mixture for chewing, a vial containing otto of roses, a *goolabpash* or bottle to sprinkle rose-water on visitors, a box for containing spices, another for holding *meesee* (a powder made of gall-nuts and vitriol for *blackening*[1] the teeth), one for *soorma* to blacken the eyelids, one for *kajul* to darken the eyelashes, a comb, a looking-glass, etc.

This *kajul* is used in the same way as the Egyptian kohl, often mentioned before, but the *soorma* is applied inside the eyelids, and there is a very curious tradition connected with the origin of this custom. They say that when God commanded Moses to ascend Koh-e-Toor (Mount Sinai), to show him His countenance, He exhibited it through an opening of the size of a needle's eye, at the sight of which Moses fell into a trance. After a couple of hours, on coming to himself, he discovered the mountain in a blaze, when he descended immediately. The mountain then addressed the Almighty thus:—"What! hast thou set me, who am the least of all mountains, on fire?" Then the Lord commanded Moses, saying, "Henceforth shalt thou and thy posterity grind the earth of this mountain, and apply it to your eyes." Since then this custom has pre-

---

[1] Women blacken their teeth when they marry, and keep them so as long as their husbands are alive.

vailed, and the *soorma* sold in the bazaars of Hindostan is supposed to be earth coming from Mount Sinai.[1]

Among other perfumes used by Indian Mussulmans may be mentioned *Abeer*, a scented powder, which is rubbed on the face and body, or sprinkled on clothes and which is made of sandal-wood, aloes, turmeric, roses, camphor, and civet; another powder called *Chiksa*, composed of mustard-seed, flour, fenugreek, cyprus, sandal-wood, patchouly, kus-kus, aniseed, camphor, benzoin, and all known spices; *Uggur-kee-buttee*, a pastille made of gum-benzoin and other odoriferous substances; and *Urgujja*, a sweet ointment composed of sandal-wood, aloes, otto of roses, and essence of jasmine. They also use a tooth-powder called *Munjun*, which is a mixture of burnt almond-shells, tobacco ashes, black pepper, and salt.

Indian women pay great attention to their hair, which is generally of a beautiful colour and length, but rather coarse. They anoint it with perfumed oil, and wear in it a profusion of jewels, the poorer class substituting glass beads for those costly ornaments. Sometimes also they decorate their heads with natural flowers, the silvery jasmine or the golden champac setting off admirably their raven tresses. The blossoms of a sort of acacia, called Sirisha, they place above their ears :—

> " Fond maids, the chosen of their hearts to please,
>   Entwine their ears with sweet Sirisha flowers,
> Whose fragrant lips attract the kiss of bees,
>   That softly murmur through the summer hours."

[1] Qanoon-e-Islam Gloss. xcv.

The hair is worn by some confined in a net, but more generally in long tresses, which are united into one in

case of mourning. The nautch-girls, or *bayadères*, wear ringlets in front and plaits at the back of the head. The accompanying illustration, from a native drawing, will convey some idea of the appearance of an Indian beauty, who might lay claim here to the same appellation, were it not for the nose-ring, which may be thought objectionable, and which must decidedly be inconvenient.

Hindú Head-dress.

In the Himalayas the hair is made up into long

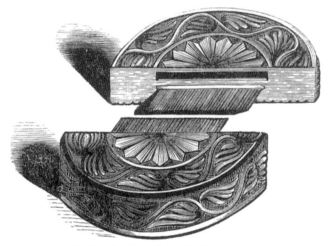

Pocket Comb used by the Mech Tribes.

braided tails, women wearing two, and men only one.

The Lepchas have, in addition, a detached braid, forming an arch of about ten inches in height over the head as represented in page 7. They pay great attention to their hair, and generally carry with them a pocket comb, curiously carved, like the accompanying specimen found amongst the Mech tribes.

We shall now proceed to the Celestial Empire, where perfumes have also been used since the earliest times.

Statue of Providence with burning Censer.

A Chinese proverb, attributed to Confucius (or Kong-Foo-Tse), says, "Incense perfumes bad smells, and candles illumine men's hearts." Acting on that principle, they use both lavishly in public and private, which would lead the hypercritical to conclude that their hearts require a great deal of lighting up, and that the natural odours of their temples and dwellings are none of the sweetest.

Joss-sticks (*wăn hëang*) and tinsel-paper (*yuen paou*) are the forms under which this incense is usually burned, and the consumption is so enormous that, according to Morrison, there are no less than ten thousand makers in the province of Canton alone. Morning and evening three sticks of incense are to be offered. They are usually placed in stationary censers of an elegant form, such as the annexed specimen taken from a temple at Tong-Choo-Foo. Sometimes they are laid at the feet of idols, as shown in the preceding illustration which represents a statue of Providence.

In the Ti-vang-mia-o, or Hall of ceremonies, at Pekin, incense is burned in twelve large urns, in memory of the deceased emperors. When the man-darins come and pay their respects to their present monarch, they also burn incense before him; if he is away they offer the same homage to his empty chair. A similar ceremony takes place every year at the festival held in honour of Confucius.

Perfumes also play their part at Chinese funerals. The body is washed, perfumed, and dressed in the best apparel of the deceased, whose portrait is placed in the middle of the room, above

Chinese Censer at Tong-Choo-Foo.

the incense-burner, which forms an indispensable item in their household furniture. The persons forming the procession who convey the corpse to its last abode burn perfumed matches all the way. The nearest relatives walk on crutches, as if entirely disabled from grief, whilst the women, carried in palanquins closed with white silk curtains, utter loud lamentations.[1]

Private Incense Burner.

[1] Lord Macartney's Embassy to China.

The catalogue of Chinese perfumery is rather limited. Besides the incense sticks, they only use a few scented oils and essences, which are more strong than agreeable— 衣 香 e *hëang*, a perfume for the clothes, and 香 革 *hëang tsaou*, a pomade for the hair. Musk is one of their favourite perfumes, which is but natural, considering that they supply all the world with it, the animal which produces it inhabiting the provinces of Mohang Mang and Mohang Vinan. They not only like its flavour, but they believe that it cures every disease under the sun, even *headache*, and in this opinion they are backed by their principal medical authorities. Pao-po-tsé recommends it as a sure preventative against the bite of serpents, and says that all persons travelling in the mountains should carry a small ball of musk under the nail of the big toe, as the musk-deer (which they call *shay*) being in the habit of eating serpents, those reptiles are kept away by the odour. Sandal-wood, patchouly, and *assafætida* complete the list of Chinese perfumery ingredients.

They have some beautifully fragrant flowers, such as the Kwei-Hwa (*Olea fragrans*), Lien-Hwa (*Nymphæa nelumbo*), Cha-Hwa (*Camellia sesanyna*), and a sort of jasmine called Mo-lu-Hwa, one blossom of which is sufficient to scent a room. They possess also several species of odoriferous woods, but they have not hitherto availed themselves of these natural treasures. They hold, however, in high esteem the fruit of a cedar which grows in the mountains of Tchong-te-foo, and hang it up in their rooms to perfume them.

Soap is not made or used by the Chinese. A natural alkali, called "keen," which is found in abundance near Pekin, serves as a substitute for washing their clothes. As to their persons, I am forced to confess that they do not appear to feel the want of a detersive, their taste for ablutions being very limited. If, however, soaps are not in request with Chinese *belles*, they have not the same objection to cosmetics, which they apply very liberally to their skin. Those who have some regard for their complexion, bedaub themselves at night with a mixture of tea-oil and rice-flour, which, like the Roman dames, they carefully scrape off in the morning. They then apply a white powder called "Meen-Fun," touch up with a little carmine their cheeks, their lips, their nostrils, and the *tip of their tongue*, and sprinkle rice-powder over their face, which finishes the elaborate picture, and softens its tones. Some of them also use the pulp of a fruit called Lung-ju-en, with which they make a sort of cold-cream for the skin.

Chinese Maid.

There are three styles principally adopted by a Chinese lady for dressing her hair, which styles indicate whether she is a maid, wife, or widow. From her infancy to her marriage, a young girl wears the back part of her hair braided into a tail, and the remainder combed over her forehead, and cut in the shape of a crescent. On her wedding-day, her head is decorated with a crown covered with tinsel paper, and on the next day her hair is dressed, for the first time, in the well-known *teapot style*, of which the annexed engraving is an illustration. On holidays she ornaments it with flowers, either natural or artificial, according to the season. When she becomes a widow, she shaves part of her head, and binds round it a fillet, fastened with numerous bodkins, which are sometimes very costly.

Chinese Head-dress.
(Teapot Style.)
[*From an original sketch by Mr. E. Greey.*]

The men shave their heads, keeping only on the summit a long tuft of hair, of which they are very proud, although it was originally a mark of their subjection to the Tartars. When their hair is thin, they mix silk or horsehair with it, to give their tails a respectable appearance. Sometimes they wind this appendage round their necks when they are at work; but if they see a stranger approaching they quickly restore it to its natural position, as it would be thought unmannerly to receive any one in that state.

Barbers are called in China *Tc tow tëïh jin*, or literally "shavers of the head," this being their principal occupation; but like the barber-surgeons of old

they combine with shaving, bleeding and other operations. They exercise their calling in the open air, as
represented in the illustration below.

We find in Japan many customs similar to those of

Chinese Barber.

the Chinese. Their list of perfumes is also rather
limited, and consists chiefly of a pomatum called Nioi-
abra, made of oil and wax; Jinko, an aromatic wood
used for burning in temples and private houses; a sort

of sachet called Nioi-bukooroo; and Hamigaki, a tooth-powder made of fine shells found on the coast, and mixed with scented herbs. European perfumes are slowly working their way into the country, but not much consumption is to be expected until *paper pocket-handkerchiefs* are abolished. Aromatics are used in funeral rites, somewhat in the same manner as they were by the ancient Greeks and Romans. The body is placed on a pile of fragrant woods, the youngest child of the deceased sets fire to it with a torch, and all persons present throw on it oil, aloes, and odoriferous gums.

Cosmetics are as much used by ladies in Japan as they

Japanese Ladies at their Toilet.
(*From the Mirror of Female Education, published at Jeddo.*)

are in Kathay; and, if we may judge by the above sketch, the duties of the toilet are an important matter with them. I have in my possession a Japanese book, from which I have selected the accompanying portrait of a *belle* in full dress, one of those charming creatures thus apostrophised by a native poet :

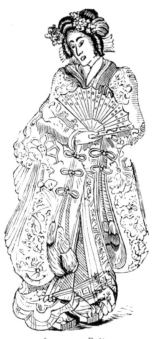

Japanese Belle.

> "One glance of her eye
> And you lose your city ;
> Another, and you would
> Forfeit a kingdom."

Japanese ladies pay great attention to their hair, which they arrange in all manner of fantastical styles, inserting into it an immense quantity of pins, made of tortoiseshell or lacquered wood, and sometimes also natural flowers. When a woman marries, she blackens her teeth and extirpates her eyebrows. The men shave the fore-part and the crown of their heads, and work up the back and side hair into a tuft over the bald skull. The annexed engraving represents the ordinary kind of female head-dress, and the illustration forming the frontispiece of this chapter, which is a perfect fac-simile of a Japanese fashion-plate, from the *Hair-Dressers' Journal*

Japanese Head-dress.

at Nagasaki, proves that both men and women indulge in a great variety of styles and ornaments. The lower part of the plate is composed of ladies' *coiffures*, and the upper part is reserved to the sterner sex, which is indicated by the blue patch on the head showing where it is shaved.

Thus they take great pains to get rid of what we are so anxious to preserve; and glory in a smooth pate, which we Europeans endeavour to conceal with a peruke. So much for diversity of tastes in nations. Some shave their heads, and others their chins, and each calls the other uncleanly for not following the same fashion!

OTAHITIAN DANCERS.

CHAPTER IX.

UNCIVILIZED NATIONS.

"Mit Perlen, die Persia's Fluth gebar
Durchflicht sie das krause, das schwarze Haar,
Schmückt die Stirne mit wallenden Federn, und
Den Hals und die Arme mit Muscheln bunt."

FREILIGRATH.

T was mentioned at the commencement of this book that civilized people would not monopolise our whole attention; but that among savage tribes we could also find some curious fashions to chronicle. In every age and in every country, men, even in a barbarous state, have attempted to enhance artificially their personal attractions; and however indifferent their success may have been in our eyes, it is only charitable to suppose that it attained its purpose with them. A Botocudo dandy, parading about with a huge wooden disc inserted in his lower lip, thinks no doubt as much of himself as one of our fops issuing in full trim from the hands of his valet; and who is to decide, after all, which is the true standard of taste? Let those who think that *we* must always be in the right look back to the fashion plates of fifty or sixty years ago, and it is highly probable they will irreverently apply the name of *old guys* to their grandfathers and grandmothers; but may we not naturally expect our grandchildren to entertain the same flattering opinion of ourselves in half a century?

Before concluding our history, therefore, and bringing it down from the Roman Empire to the present time, we shall devote this chapter to a glance into various nooks and corners of the world where, although they know little or nothing of civilisation, they still attempt to ornament and decorate in various ways "the human face divine." Of perfumes, properly speaking, there is a very limited use among these people, whose untutored olfactories are sometimes apt to prefer a strong rancid smell to the finest productions of our laboratories; but if we are allowed to class among cosmetics the various pigments used by them for painting their faces and bodies, we shall find them extensively patronised. And why should not the elaborate and motley colours applied by the Red Indian to his physiognomy, to render him by turns attractive to his squaws or terrible to his foes, be placed in the same category with the *patent enamel* of some of our London aspiring *belles*, who confidently believe it will make them *beautiful for ever?* As to the modes of dressing the hair, numerous and eccentric as may be our styles of European coiffures, they are left altogether in the shade, when compared with the extraordinary contrivances resorted to by the children of nature in decking out the hair or *wool* which may have fallen to their lot.

Commencing our tour in Africa, we shall find the custom of anointing as prevalent with all the natives as it was with the ancient Greeks and Romans, and applied, as was the case then, to the body as well as to the hair. The chief motive for this practice is no doubt a sanitary one; by means of this greasy coating they protect their skin against the scorching rays of the sun, on the same principle that a cook bastes her meat well to prevent it from burning; but it is also looked upon by them as a great embellishment. They take as much pride in exhibiting a sleek, oily cuticle as a Parisian in wearing well-polished boots, and no greater compliment can be paid to a woman than to say she looks "fat and shining." They accomplish this desirable result by means of various lubricating substances, such as cocoa-nut oil, palm-oil, and a kind of butter called *ce*, produced by pounding in a mortar and boiling in water the fruit of a tree which grows on the west coast of Africa. These ointments are generally flavoured with aromatic herbs or scented woods; but from the accounts of travellers, their aroma

is often "more peculiar than pleasing." That it is
strong enough is not to be doubted, for Mr. Hutchinson,
in his "Ten Years in Æthiopia," speaking of a particular sort called Tola pomatum, which is used in the

A Bridegroom's Toilet at Fernando Po.

province of Fernando Po, says, "The first thing of
which one is sensible when approaching a village is the
odour of Tola pomatum, wafted by whatever little breeze
may be able to find its way through the dense bushes."

The same traveller gives the following amusing

account of the "toilet" of a Fernandian bridegroom :—
"Outside a small hut, belonging to the mother of the
bride expectant, I soon recognized the happy bridegroom undergoing his toilet from the hands of his
future wife's sister. A profusion of Tshibbu strings
being fastened round his body, as well as his legs and
arms, the anointing lady, having a short black pipe in
her mouth, proceeded to putty him over with Tola
paste. He seemed not altogether joyous at the antici-
pation of his approaching happiness, but turned a sulky
gaze now and then to a kidney-shaped piece of yam
which he held in his hand, and which had a parrot's
red feather fixed on its convex side. This, I was
informed, was called Ntshoba, and is regarded as a
protection against evil influence on the important
day."

It must not be supposed that this beautifying process is confined to the male sex ; for, speaking a little
further on of the bride, Mr. Hutchinson says—"Borne
down by the weight of rings and wreaths, and girdles
of Tshibbu, the Tola pomatum gave her the appearance
of an exhumed mummy, save her face, which was all
white, not from excess of modesty (and here I may add
the negro race are reported always to blush *blue*), but
from being smeared over with a white paste, the symbol
of purity. As soon as she was outside the paling, her
bridal attire was proceeded with, and *the whole body
plastered over with white stuff.*" What a pretty substitute
for the classical wreath of orange-blossoms, and what a
charming contrast must be offered when the paint

gradually peels off, and reveals the sable ground on which it is laid !

Dr. Livingstone, Du Chaillu, and other African explorers give us amusing accounts of the fantastical

Bushukulompo Head-dresses.

modes of native hair—or, rather, *wool*—dressing. The Bushukulompos work theirs up into a cone somewhat like a helmet,[1] whilst the Londa *ladies*[2] bring theirs over

Londa Head-dress.

in front and at the back of the head, in the shape of a cocked hat, with a carved pin jauntily stuck in, in lieu of a feather. The Ashira *belles* patronise a more elaborate style, consisting in multitudinous points radiating from the face, and confined with

an outward circle, which would give them some faint

[1] Dr. Livingstone's Africa.   [2] Du Chaillu's Travels.

resemblance to a saint such as depicted in Catholic countries, if the picture contained inside the nimbus wore a more angelic expression. The Makololo women cut their hair quite short, and in the Great Desert of Sahara the forehead is shaved high up, leaving only one curl, which is braided and hangs down over the face.[1] The Hottentots, according to Sir John Barrow, have very curious hair; it does not cover the whole surface of the scalp, but grows in small tufts separated from each other, and, when kept short, looks and feels like a *hard shoe-brush*.

Ashira Head-dress

The most varied and extraordinary *coiffures*, however, are to be found among the tribes of the Ounyamonezi, or Mountains of the Moon, as will be shown by the group, on the next page, taken from Capt. Burton's interesting Voyage to the Lake Regions of Central Africa. To complete their attractions they have two deep scars made on each side of the face with a razor or a knife. This ornament is also patronized by the fair sex; but with their usual *penchant* to coquetry, they have the scars dyed of a blue colour.

In the island of Madagascar, the long black hair of the men used to be plaited in small tails, three or four inches in length, with a knot at the end; but King Radama, finding this fashion inconvenient for his troops,

[1] Richardson's Travels in the Great Desert of Sahara.

published an edict ordering all his soldiers to have these
plaits cut off. This law, however, met with great oppo-
sition, not only from the men—who cherished their
capillary ornaments as much as the hussars of the last
century did their tresses and queues—but also on the part
of their wives, who prided themselves on their attention

Head-dress of the Ounyamonezi Tribes.

in keeping their husbands' hair well plaited and greased
with cocoa-nut oil. Finding ordinary legal means in-
sufficient, King Radama resorted to the force of example,
and appeared one day at a review with his hair cropped
quite close. Those who were most anxious to please
their sovereign, did not now hesitate to sacrifice their
locks; but some of the more obstinate held out, en-

couraged in their resistance by the women, who raised quite an *émeute* about it. Seeing this, the king quietly instructed his guards to take the disobedient to a neighbouring wood, and cut off their hair *in such a way that it should not grew again.* The intelligent servants, with a zeal worthy of such a master, punctually obeyed these orders, for they cut off—*their heads!*[1]

The mode of plaiting the hair seems the most prevalent in Africa; for, according to Consul Petherick, we find it adopted, with a few exceptions, by both sexes over all the eastern part of that continent, from Mount Sinai to the White Nile. Respecting the Hassanyeh Arabs, who inhabit the latter locality, he says—"The heads of men and women are dressed with equal care, the hair of both being plaited, although not in a similar manner, that of the man being drawn off the forehead towards the back of the head, around which it hangs in numerous plaits. The woman collects the plaits together in bunches at each side of her face, and at the back of her head, ornamenting them with coral, amber beads, and little brass trinkets. Brass thimbles, perforated through the top, and strung on a stout thread, sustained by knots at regular distances above each other, and suspended to the crown of the head, hanging down at the back of it, form a very favourite ornament, as also does an old button or any little brass trinket over the forehead.[2]

---

[1] Captain Owen's Voyage to Africa.
[2] Egypt, the Soudan, and Central Africa, by John Petherick.

In Nubia, the hair, which is inclined to be woolly, is plaited into a variety of forms, but generally close to the head, fitting like a skull-cap, and hanging down in thick masses of innumerable small plaits all round the back and sides of the head. Another style is to plait only the part next to the head, and have the ends combed out and stiffened with a gummy solution, forming a thick bushy circle round the head. This is a very elaborate sort of *coiffure*, which is only done once or twice a month, as it takes a long time to build up; and those who patronise it are obliged to sleep with their head reclining on a small wooden stool, hollowed

Abyssinian Lady.

out to fit the neck, so as not to disarrange the precious edifice, which shows that victims to fashion are to be found even in those remote parts.[1]

Abyssinian ladies wear in their hair ivory or wooden pins and combs, neatly carved in various patterns, and stained with henna. They also indulge in a profusion of chaplets on their heads and round their necks, and the most elegant carry on their bosom a large flat silver case containing scented cotton, which they consider as a sort of amulet.

Abyssinian Amulet.

The Bedouin Arabs of Mount Sinai have their hair plaited, and so arranged as to form a protuberance resembling a horn placed low down

[1] Egypt, the Soudan, and Central Africa, by John Petherick.

on the forehead and projecting two or three inches. The girls wear on their heads a wreath of various coloured beads, to which are suspended neatly-carved oyster shells, the latter being considered as a significant hint to the young men of the tribe that they have no objection to alter their condition. This may not be quite so poetical as the language of flowers, but still it is a great pity a similar custom is not adopted in England, as the sight of the oyster-shell would naturally encourage timid young men to "pop the question."

Abyssinian Combs.

In Upper Egypt, Arab perfumery and cosmetics are extensively patronised by those who can afford them. Musk, for scenting the clothes, and kohl, to darken the eyelashes, are two indispensable items in the list of presents sent to a bride by her intended; and the latter, with a praiseworthy regard for the future wants of the community, during a few days after the marriage squats on a mat at the door of the mosque, exhibiting his presents on a tray, and collecting alms from the faithful.

Going farther into the interior, the principal article of perfumery (if it may be so called) we find in use, is a sort of pomatum or butter, more or less scented, which the natives generally keep in ostriches' eggs, and use profusely, the most stylish thing being to put a pat on

the head, and let it melt and run down the whole body.
Others apply this ointment to their heads or persons
with an ostrich's feather, which they carry about
in a case made of a buffalo-horn.   The specimen
of this singular toilet implement represented here, I
found in Mr. S. H. Berthoud's unique collection, and
is, I believe, the first that has been seen in Europe.

African Anointing Feather.

There is a very curious sort of bath used in Nubia
which deserves particular description.   Consul Petherick relates that, having ordered a *bath* at Berbera, one
of the Nubian towns he visited, he was much surprised at
seeing a negro maid enter bearing a bowl and a teacup
as the sole apparatus required.   The bowl contained
dough, and the cup a small quantity of sweet oil scented
with aromatic roots; the former of these well rubbed
on the bare skin cleaned it thoroughly, after which the
perfumed oil was applied, to give elasticity to the limbs.
The whole operation, which is called *dilka*, is in great
favour with the natives; and Mr. Petherick, who de-
clares he was much refreshed by it, attributes to its use
the entire absence of cutaneous diseases among these
people, and says it enables them to resist the cold and

cutting winds of winter with no other protection than very thin clothing.

An aromatic fumigation replaces, in the Soudan, even this very imperfect mode of bathing. In a hole, dug in the ground by the side of the bed, is placed an earthen pot, in which is burned the odoriferous wood of the tulloch. The natives sit over this, covering themselves closely with a thick woollen wrapper, and remain exposed for about ten minutes to the cloud of fragrant smoke, which causes intense perspiration, and is supposed to exercise a tonic and beneficial influence on the skin. *Ladies* who use this frequently become incrusted in time with an odoriferous enamel which is highly prized and considered *very fast*.

Even in the remotest wilds of Central Africa we find people endeavouring to assist nature with art as far as is compatible with their primitive minds.

The *Neam Nam*, a tribe in the far interior on the equator, take great pains with their hair, which they wear plaited in thick masses covering the neck, and which they ornament with long ivory pins from six inches to a foot in length. These pins are carved in pretty patterns, and partly dyed with the decoction of a root; they are inserted at the back of the head, long ones alternating with short ones, and forming a semicircle, somewhat similar to that worn by the peasant girls on the borders of the Lake of Como, the only difference being that the Italian decoration is composed of steel and gold pins. This is certainly a very curious coincidence.

The Dinkas dye their hair red, whilst the Djibbas, who are a warlike people, pride themselves in interweaving the hair of their fallen enemies with their own, forming a thick tail, the length of which indicates the valour of the wearer.

The greatest dandies, however, are the Griquas, who smear themselves with grease and red ochre, whilst the head is anointed with a *blue* pomatum made of mica. The particles of shining mica falling on the body are thought highly ornamental, and the mixture of colours very attractive.[1]

Taking a bold stride thence to the Philippine Islands, we find the natives, who are called Tagals, pay the greatest attention to their hair, which is long, black, and glossy. The women wash it at least once a day with a saponaceous grass called *go-go*, and anoint it with cocoa-nut oil scented with the flowers of the *alangilan* or *san-paquita*.

Both men and women in the Loo-Choo Islands wear their hair drawn up towards the crown, and worked up into a sort of loop, which is ornamented with two pins. The wealthy have these pins studded with precious stones, and use the juice of an aromatic plant to enhance the natural brilliancy of their hair.

Javanese women greatly pride themselves on the yellow complexions which nature has allotted to them. It is the constant theme of their poets, who praise its *golden* hue with as much fervour as ours do the roses and lilies which distinguish our *belles*. Admiral Du-

[1] Dr. Livingstone's Africa.

mont d'Urville says they have recourse to yellow cosmetics to keep up the brilliancy of the favourite tint,
in the same way as rouge and white are used here.[1] In
addition to this they blacken their teeth, and greatly
ridicule Europeans for the whiteness of theirs which,
according to their opinion, makes them look like *monkeys*.

In Australia the aborigines are worse than Esqui-
maux: to these tribes a bad smell is really a perfume,
so we will leave them alone. Yet the country produces
plenty of sweet-scented flowers and plants, and whole
forests of trees with fragrant leaves;[2] and who knows
but one day that fertile market for our manufactures
may in its turn furnish the world with essences and
cosmetics? When, in a few centuries, Lord Macaulay's
New Zealander takes his stand on the ruins of London
Bridge, his handkerchief made of the fibres of the *for-
mium tenax* will probably be redolent with the *last new
scent* by Warranonga of the Murrumbidgee!

Tattooing ranks among the chief personal adornments
with the Australian and Polynesian races. It might
almost be called an indelible form of cosmetic, for it
probably originated in facial painting; some savage of
enduring cuticle having conceived the idea of rendering
the colour permanent by driving it into the skin. New
Zealand bears, or used to bear, the palm in this art.
There the chiefs especially prided themselves on the elegant arabesques which decorated their physiognomies,
and hair and beard were willingly sacrificed to afford a

---

[1] Voyages autour du monde, par Dumont D'Urville, vol. ii. p. 324.
[2] Principally the Eucalyptus and Melaleuca tribes.

better ground for the design. This operation, called
moko, was generally performed with a black powder
composed of the burnt resin of the *kauri*, which was
inserted into the skin by means of a small chisel made
of the bone of an albatross. The process is described at
full length by Mr. Taylor in his interesting work on
New Zealand. He says that to allay the pain caused
by it the artist used to sing to his patients songs, of
which he gives the following curious specimen :—

> He who pays well let him be
> Beautifully ornamented ;
> But he who forgets the operator
> Let him be done carelessly :
> Be the lines wide apart.
>   O hiki Tang iroa !
> Strike that the chisel as it cuts
> Along may sound :
> Men do not know the skill of the operator
> In driving his chisel along.
>   O hiki Tangaroa !"

The delicate allusion contained in these lines shows that
artists in their poetical effusions always had "an eye to
business," and thought it necessary to remind their
patrons that *beauty*, like everything else, must be paid
for. The gentler sex had likewise recourse to this mode
of embellishment, but the tattooing was only executed
on the lips and chin, with an arch little curl at the
corner of the eye by way of an *accroche cœur*.

Embalming seems also to have been practised by New
Zealanders, but was confined to the heads of cherished
relations, which, after taking out the brain, were stuffed
with flowers, baked in ovens, and finally dried in the
sun. These heads were kept in baskets carefully made

and scented with oil. They were brought out on grand occasions, ornamented with feathers, and cried over by all the family.

The most extraordinary and fantastical coiffures are perhaps to be found among the Feejee Islanders. Not satisfied with twisting their locks into every conceivable shape, they vary their sable appearance by dyeing them in sundry colours, such as blue, white, red, and yellow. Among young people bright crimson and flaxen are the favourite hues; but the most fashionable style is to combine several shades in the same head-dress. Thus some wear a spherical mass of jet-black hair with a white band in front as broad as the hand; or a white oblong occupies the length of the head, the black hair passing down on either side; whilst others have a large red roll or a sandy projection falling on the neck; and others, again, work fancy devices on their hair, dividing it into squares or cones of different hues. I humbly submit this *notion* to ladies fond of novelties, and am certain that such a chequered head-dress would create quite a sensation in one of our drawing-rooms. Hair-dyes we are well acquainted with; and that some of them are apt to produce varied shades, from a lively pea-green to a soft violet, is no secret to those who use them: but, with our anti-Feejean prejudices, we have considered this circumstance hitherto rather as a misfortune than a matter of ornament.

The natives of Duke of York's Island are also partial to hair of divers hues; but they attain their purpose without dyeing it, by simply smearing it with grease and sprinkling it afterwards with a white, red, or yellow powder made of burnt shells and coral, which they carry always with them in a small gourd—a fashion which

Marquesas Hair-pin.

reminds one of our own powdered beaux of the last century.

Several other tribes of South Sea Islanders patronize

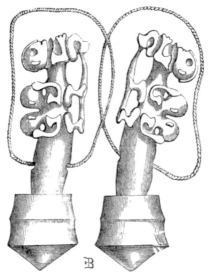

Marquesas Ear-rings made of Fish-bones (natural size)

multi-coloured hair, and among others those of the Darnley and Britannia Islands. Young people of the latter group take as much trouble in whitening their black locks as elderly Europeans in blackening their white ones.

At Nooka-hiva, the principal of the Marquesas Islands, both sexes anoint themselves freely with sweetly-scented cocoa-nut oil, and the most refined use as a substitute the juice of the *papa*, which is supposed to whiten the skin and preserve its smoothness. The women bestow particular care on their hair, which they ornament with

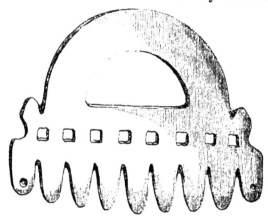

Comb from the Solomon Islands.

long carved pins: they also wear ear-rings, generally formed of fish bones. The accompanying specimens are from Mr. Berthoud's collection, as well as the comb from the Solomon Islands, which is made of the teeth of the sea-elephant (*Trichechus*).

Last, not least, we must mention Tahiti, the Queen of the Pacific, where the natives, and especially the women, have always paid great attention to their personal appearance. Since their contact with Europeans they have adopted many of their customs, and they are not now the same as described by Captain Cook and as represented in the frontispiece to this chapter; but still they have preserved some of their original habits, which are worth noticing.

The Tahitian women are generally tall and well made; they have fine eyes and teeth, and beautiful long hair, to which they devote great care. They wash it daily, anoint it with a pomatum called *monoï*, made of cocoa-nut oil, scented with sandal-wood or *toromeo* root, and plait it in long braids, which hang down their back. Sometimes they work it up into a sort of diadem, ornamented with odoriferous flowers called *maïri*, or with the deliciously-scented blossoms of the *Tiare*, a sort of jasmine. The *reva-reva*, formed of cocoa-nut tree fibre, is another favourite head-dress with them, and very elegant crowns are also made with the arrow-root straw or *pia*. Specimens of some of these *coiffures* are exhibited at the Colonial Museum in Paris, and are extremely graceful.

Crossing over now to America, and commencing with the southern extremity, we find a curious custom recorded by Captain Cook as existing then in Terra del Fuego, and in all probability the same is still in vogue. The natives of that country paint themselves all over with red and white, the red forming patches on the chest and shoulders, and the white long streaks on the arms and legs. With a little white round the eyes, and a long bone passed through the cartilage of the nose, their toilet is considered complete.

The South American Indians generally have long

black hair, which they wear loose on their shoulders. The women plait theirs behind with a ribbon, and cut it in front a little above the eyebrows from one ear to the other. The greatest disgrace that can be inflicted upon Indians of either sex is to cut off their hair; they will put up with any corporal punishment in preference, and such a measure is consequently limited to the most enormous crimes. They are nearly all very fond of perfumes, but, although their soil abounds in aromatic materials, they generally resort to our European productions. There is, however, a native perfume mentioned by Mr. Wallace as being very exquisite and in great repute on the Rio Negro. It is called *umari*, and is extracted from the *humirium floribundum* by means of a very singular process; this consists in lifting the bark and inserting under it pieces of cotton wool to imbibe gradually the scent which is expressed from them at the end of a month.[1]

We shall now conclude our long ramble with the North American Indians, and briefly describe their mode of face-painting, an art in which they certainly are unrivalled. From all accounts of travellers who have visited the *Redskins*, no dowager of the *ancien régime*, rougeing and patching for the opera or ball, ever spent so much time at her toilet as a Sioux or a Pawnee *getting his face up* for an excursion either of a warlike or a peaceful nature.

Mr. Murray, speaking of the son of a chief called Sa-in-tsa-rish, says that he never saw any dandy to equal

---

[1] Travels on the Amazon and Rio Negro, by A. R. Wallace.

him for vanity. He usually commenced his toilet at eight o'clock in the morning, and it was not concluded until a late hour; after having greased his whole person with fat to serve as a ground for the paint, and drawn a few streaks on his head and body, he kept looking at himself in a bit of mirror he carried with him, and altering the lines until they happened to please him.

Some pretend that there is a certain symbolism in the various colours they use; thus, for instance, red typifies joy, and black mourning. In this latter particular they exhibit some resemblance to ourselves, the difference being that, instead of assuming a sable garb when they lose a relative, they rub their face over with charcoal. The subdued tints of half-mourning they represent with a trellis-work of black-lines over the face, or sometimes they paint one-half of their face black, as we do the ground of our family escutcheons. Fortunately they are not addicted to frequent ablutions, or their mourning would be of short duration.

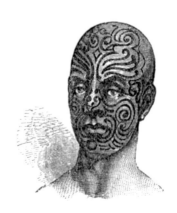

A STROLLING VENDOR OF PERFUMERY. (Time of Louis XV.)

CHAPTER X.

From Ancient to Modern Times.

Cerca d'accrescer collo studio e l'arte
La natural beltá che in lei risplende,
L'auree chiome in vago ordine comparte,
Ed ad ornarsi il rimanente attende :
Poi lieta si contempla a parte a parte
Nell'acciar, che l'immago al vivo rende,
Così augellin dopo la pioggia al Sole
Polirsi i vanni, e vagheggiarsi suole.

TORQUATO TASSO.

EAVING far distant lands, we shall now return to our own Europe, and trace from the earliest times the progress of the art which forms our subject, principally in England, France, and Italy, with respect to which countries our information is most complete.

The toilet of the ancient inhabitants of Britain somewhat resembled that of the North American Indians, and consisted in a series of elaborate paintings on the whole surface of the body, which was no doubt originally intended to protect the skin against the inclemencies of the weather, but which was afterwards used as a mode of embellishment and a means of distinguishing the different conditions; for it was reserved to freemen, and strictly forbidden to slaves.[1] The common people only indulged in small designs, drawn at a distance from each other, whilst the nobility had the privilege of ornamenting their persons with large figures, chiefly of animals, which were subsequently transferred to their shields when they adopted a less scanty costume. This may be looked upon as the origin of family arms, which the Japanese, who probably commenced in the same way, now wear embroidered on their dress.

The Picts who inhabited the North of Britain, were the most remarkable for their pictorial decorations, whence they derived their name.[2] The Gauls and the Germans dyed their breasts red before going to fight, so that the enemy could not see the blood flowing from their wounds. Among the various colouring substances then in use, Julius Cæsar mentions *woad* (*Isatis Tinctoria*), with which the Britons gave a bluish cast to their skins, and made themselves look dreadful in battle.

[1] Pelautier, " Histoire des Celtes."   [2] *Picti*, "painted."

Pliny also speaks of a sort of plantain called *Glastrum*, by means of which the Gauls and Britons stained their faces and bodies.[1]

Hair-dyes were already known even at that early period, for Diodorus Siculus says that the Britons, who naturally possessed red hair, endeavoured all they could to make it redder by art, which they accomplished by washing it repeatedly in water boiled with lime.

The Druids left no written record of their customs, but, from contemporaneous accounts, they do not appear to have used perfumes in their mode of worship, which was of the most primitive description. They knew, however, and highly prized, the numerous aromatic plants indigenous to their soil. Druidesses crowned their brows with verbena, and composed with fragrant herbs mysterious balms, which cured the heroes' wounds and enhanced the charms of the fair.

The Roman conquest brought into Gaul and Britain the civilised manners of the conquerors. Body painting and rude ornaments were laid aside and exchanged for graceful costumes and elaborate cosmetics, and the provinces soon equalled the metropolis in elegance and refinement. The various toilet implements and splendid baths of that epoch, discovered by excavations in France and in England, bear witness to the high state of luxury which existed then in those countries. This, however, lasted but a time, and with the Roman dominion ended this transient gleam, for all relapsed into darkness with successive invasions.

[1] Pliny's Nat. Hist. lxxii. cap 1.

From that period to the Crusades, the principal records of perfumes we find in history are connected with
the church or the court, for they were then too costly
to be used much in private life. In the year 496, when
Clovis, the first Christian king of France, was baptised
at Rheims, incense was burned, and fragrant tapers
were lighted, for that ceremony.

That incense was also known to the Anglo-Saxons
appears by the following riddle translated from the
Exeter Book :—

> " I am much sweeter than incense or the rose
>   That so pleasantly on the earth's turf grows ;
>   More delicate am I than the lily,
>   Though dear to mankind that flower may be." [1]

Hugh the Great, father of Hugh Capet, having asked
in marriage the sister of King Athelstan, sent, among
other presents, as the Malmesbury Chronicles inform
us, such perfumes as had never been seen in England.
Charlemagne was also a great lover of scents, and at
his brilliant court at Aix-la-Chapelle they were in
constant request.

Carpets were not known then, but they used to strew
on the floor, in the houses of the great, sweet rushes,[2]
which spread a pleasant fragrance through the atmosphere. When William the Conqueror was born in
Normandy, where that custom prevailed, at the very
moment when the infant burst into life and touched
the ground, he filled both hands with the rushes on the
floor, firmly grasping what he had taken up. This was

[1] Exeter Book, p. 423.   [2] Probably the *calamus aromaticus*.

hailed as a propitious omen, and the persons present declared the boy would be a king.[1] This custom of strewing sweet rushes was still in vogue in England during the reign of Queen Elizabeth, and Shakspeare frequently alludes to it in his plays.

Embalming was sometimes practised in those days, and in Eadmer's life of St. Anselm we find the body of the saint was anointed with balsam after his death.[2]

After the Crusades, perfumes came into more general use. The gallant knights brought home to their lady loves some of the far-famed perfumes of the East, and specimens of the wonderful cosmetics by means of which the beauties of the harem preserved their charms; and among the costly presents offered to St. Louis, King of France, rare and precious aromatics formed a conspicuous part. Rose-water was also introduced about that time, and it became the custom to offer it to guests in noblemen's houses to wash their hands with after meals —a very necessary ablution, if we consider that forks, which were invented in Italy during the fifteenth century, were not known in England until the reign of James I., and were then considered a great piece of foppery. Matilda, queen of Henry I., received from France, as a present, a beautiful silver peacock, with a train set in pearls and precious stones, which was intended to contain rose-water and to be placed on the table for the above-mentioned purpose. Mathieu de Coucy also relates in his Chronicles, that, at a grand banquet given by Philip the Good, Duke of Burgundy, there

---

[1] Reliques in Malmesbury, p. 299.   [2] Eadmer, Vita S. Anselmi, p. 893.

stood on the sideboard the statue of a child, from which issued a jet of rose-water.

Perfumers had already sprung into commercial existence in France in the twelfth century, for Philip Augustus granted them in the year 1190 a charter, which was confirmed by John in 1357, and afterwards by Henry III., in 1582. That charter was for the last time renewed and enlarged by Louis XIV. in 1658. It was then requisite to serve four years as apprentice, and three years as companion, to be elected master perfumer, which shows that it was considered a handicraft of some importance.

A Lady at her Toilet (13th century).

In a manuscript of the thirteeenth century, preserved in the British Museum,[1] we find the annexed illustration of a lady at her toilet, which may convey some idea of the manner in which those duties were performed. Early morn was the time chosen for that important task by the fair of the period, as we read in the romance of " Alisaunder :"—

> " In a moretyde[2] hit was,
> Theo dropes hongyn on the gras;
> Theo maydenes lokyn in the glas
> For to tyffen[3] heare fas."

The moralists and satirists of that age reproach the ladies with paying too much attention to their personal

[1] MS. Addit., No. 10,293, fol. 266.　　[2] Morning.
[3] Adorn, from the French *attiffer*.

embellishments, and with deforming their bodies with stays, which were introduced about that time. They are also accused of painting their faces, dyeing their locks, and plucking out superfluous hair.

Our ancestors were very fond of flowers, which they used to decorate their persons as well as to ornament their gardens. Like the ancient Greeks and Romans, they wore on their heads, at all their entertainments, wreaths of flowers called in the French of that period, *chapels* or *capiels*. In the Romance of "Perce-Forest," the author, describing a festival, says, "avoist chascun et chascune un chapel de roses sur son chief."[1] These words clearly denote that even the *dark* sex indulged in this floral head-gear, which, when coupled with a rubicund nose and a

<blockquote>"Fair round belly, with good capon lined,"</blockquote>

must have produced a very pretty effect.

Ladies making Garlands,

The task of culling flowers for garlands was generally entrusted to ladies, and the above engraving, taken

---

[1] Every man and every woman had rose-wreaths on their heads.

from a manuscript in the British Museum,[1] represents
them engaged in this charming occupation. Thus is
Emelie described by Chaucer in his "Knight's Tale :"

> "Hire yolwe heer was browdid [2] in a tresse,
> Byhynde hire bak, a yerde long, I gesse.
> And in the gardyn at the sonne upriste,[3]
> Sche walketh up and doun wheer as hire liste;
> Sche gadereth floures, partye whyte and reed,
> To make a certeyn gerland for hire heede."

Jean de Dammartin, in "Blonde of Oxford," finds likewise his mistress in a meadow making flower wreaths.

> "A dont de la chambre j'avance
> De là le vit en i-prael
> U ele faisoit un capiel." [4]

Perfumery did not form then a separate branch of
trade in England. It was generally sold by mercers,
who also combined with that trade the sale of a
variety of toilet implements, such as combs, mirrors,
fillets for the head, etc. We find them mentioned in
a very curious manuscript entitled the "Pilgrim,"[5]
wherein a lady who keeps a mercery shop thus
enumerates the different articles in which she deals :—

> "Quod[6] sche, 'Geve[7] I schal the telle,
> Mercerye I have to selle ;
> In boystes sootè oynementis [8]
> Therewith to don allegementis ;[9]
> I have knyves, phylletys, callys,
> At ffeestes to hang upon wallys ;
> Kombes mo than nyne or ten,
> Both ffor horse and eke for men ;
> Merours also, large and brode,
> And ffor the syght wonder gode.' "

---

[1] M.S. Reg. 2 B. vii.   [2] Her yellow hair was braided.   [3] At sunrise.
[4] Advancing from the room, I see her in a meadow making a chaplet.
[5] MS. Cotton, Tiberius A. vii.   [6] Said.   [7] If.
[8] In boxes sweet ointments.
[9] Give relief, from the French *donner allègement.*

The accompanying cut, taken from the same manuscript, represents the mercer's shop, with some of the articles described.   The fair trader is offering to the pilgrim a flattering mirror, in which people see themselves handsomer than they are, but it is indignantly rejected by the pious man.

A Mediæval Perfumer's Shop.

Alcoholic perfumes do not appear to have been known until the fourteenth century, and the first we find mentioned is Hungary water, so called because it was first prepared in the year 1370 by Queen Elizabeth of Hungary, who had the recipe from a hermit, and became so beautiful through the use of it, that her hand was asked in marriage at the age of seventy-two by the king of Poland. This story, which is taken from an old book published at Frankfort in 1639, is related by Beckmann,[1] who devotes a whole chapter to the subject, but ends by doubting its accuracy—a most ungallant conclusion, for he ought not to question the captivating powers possessed by ladies of any age, with or without the aid of *Hungary water*.

The fifteenth century, that brilliant *cinque cento*, of which Italy is justly proud, saw the revival of the fine arts on that classical ground. The palaces of its princely merchants teemed with luxuries of every description, among which perfumery was, as usual, called to play its part. Venice, from its early intercourse with Constantinople, was one of the first to introduce the fragrant treasures of the East. In the course of time cosmetics were also adopted by its patrician dames, who, not content with the charms which nature had lavished upon them, sought to enhance them by artificial means. The first book on this subject appeared in the sixteenth century, under the auspices of Countess Nani,[2] and contained many curious recipes, among which were some for dyeing the hair of that beautiful shade called *capelli fila d'oro*.[3] As my fair readers may wish to know how this was accomplished, I shall mention one of these preparations, which consisted of two pounds of alum, six ounces of black sulphur, and four ounces of honey, distilled together with water. Cesare Vecellio, the cousin of Titian, in his interesting work, *Degli habiti antichi e moderni*, explains how this water was applied. Ladies repaired to the terraces on the tops of their

---

[1] Beckmann's History of Inventions, vol. i. p. 315.
[2] Ricettario della Contessa Nani.
[3] Golden thread hair.

houses, soaked their hair well with the preparation, and remained sitting there for hours, to let the sun well fix the colour in. They wore on their heads a large straw hat without a crown, called *solana*, to protect their complexions, and allowed their hair to hang round over

La donna che si fa biondi i capelli.

the rim until it was completely dry. The above illustration, copied from his book, will show how it was done. It is generally supposed that those beautiful

golden locks which are so much admired in the paintings of the Venetian artists of the period were acquired
in that manner, for they are seldom to be met with
among the modern population.

When Catherine de Medicis came to France to marry
Henry II., she brought with her a Florentine named
René, who was very expert in preparing perfumes and
cosmetics. His shop on Pont au Change became the
rendezvous for the *beaux* and *belles* of the period,
and from that time perfumery came into general use
among the wealthy. This René also possessed the art
of preparing subtle poisons, and his royal mistress is
said to have had frequent recourse to his talents to get
rid of her enemies. Among her victims the historians
mention Jeanne d'Albret, mother of Henry IV., and
state that she was poisoned by wearing some perfumed
gloves presented to her by Catherine; but modern
chemists doubt whether it was possible to poison any
one by such means.

In public festivals it became the custom to perfume
fountains; and in the year 1548 the city of Paris paid
the sum of six golden crowns to Georges Marteau
"pour herbes et plantes de senteur pour embaumer les
eaux des fontaines publiques lors des derniers esbatte-
ments." [1]

Under the reign of that effeminate monarch, Henry
III., the abuse of perfumes became so great that it was
denounced by the satirists of the period; and, among

[1] For aromatic herbs and plants, to perfume the waters of public fountains during the late rejoicings.

others, Nicolas de Montaut, in his "Miroir des François" (1582) reproaches ladies with using "all sorts of perfumes, cordial waters, civet, musk, ambergris, and other precious aromatics to perfume their clothes and linen, and even their whole bodies."

The earliest French perfumery book that I have met with is entitled "Les secrets de Maistre Alexys le Piedmontois,"[1] and contains some curious recipes for making pomatum with apples,[2] pomanders against the plague, "oiselets odoriférants" for burning in apartments, paste for perfuming gloves, and various hair dyes and cosmetics. To give some idea of the state of the art at that period, I shall quote the following formula for preparing a marvellous water, warranted to make ladies "beautiful for ever."

"Take a young raven from the nest, feed it on hard eggs for forty days, kill it, and distil it with myrtle leaves, talc, and almond oil."

This is a fair specimen of the whole, which strongly savours of the still prevalent delusions of alchemy, and bears no little resemblance to the recipes quoted in chapter VIII. as being still used by the Arabs.

Perfumes did not come into general use in England until the reign of Queen Elizabeth. Howes, who continued Stowe's chronicle, tells us that they could not make any costly wash or perfume in this country until about the fourteenth or fifteenth year of the queen,

---

[1] The secrets of Master Alexis, the Piedmontese.

[2] Pomatum was first prepared from apples, whence it derives its name.

when the Right Honourable Edward de Vere, Earl of Oxford, came from Italy and brought with him gloves, sweet bags, a perfumed leather jerkin, and other *pleasant things;* and that year the queen had a pair of perfumed gloves, trimmed only with four tufts or rows of coloured silk. She took such pleasure in these gloves that she was pictured with them upon her hands, and for many years afterwards it was called the "Earl of Oxford's perfume." On another occasion, Queen Elizabeth, visiting the University of Cambridge, was presented with a pair of perfumed gloves, and was so delighted with them that she put them on at once. She also usually carried with her a pomander (or *pomme d'ambre*), which was a ball composed of ambergris, benzoin, and other perfumes; and she was once mightily pleased with the gift of a "faire gyrdle of pomander," which was a series of pomanders strung together and worn round the neck. These pomanders were held in the hand, to smell occasionally, and were supposed to be preservatives from infection. They were very generally used, as may be seen from the portraits of the period. Their exact ingredients are thus described in an old play :—"Your only way to make a good pomander is this : Take an ounce of the finest garden mould, cleaned and steeped seven days in change of rose-water; then take the best labdanum, benzoin, both storaxes, ambergris, civet, and musk; incorporate them together, and work them into what form you please. This, if your breath be not too valiant, will make you smell as sweet as any lady's dog."

Drayton, in his "Queen of Cynthia," also alludes to pomanders in the following lines :—

> "And when she from the water came,
> When first she touched the mould,
> In balls the people made the same
> For pomanders, and sold."

Some of these pomanders consisted in globular vessels containing strong perfume, and perforated with small holes, not unlike our modern pocket cassolettes. The earliest illustration of this favourite toilet requisite occurs in the "Boat of Foolish Women,"[1] a series of five caricatures published by Jodocus Badius in 1502, and intended to flagellate the abuse made of the five

The Boat of Foolish Smells.

senses. The above engraving represents the "Boat of Foolish Smells,"[2] in which are three ladies, one of whom is holding some flowers she has gathered, and smelling at the same time a pomander which her friend has bought from an itinerant vendor of perfumes.

[2] Scaphæ Fatuarum Mulierum.   [1] Scapha olfactionis stultæ.

The principal perfumes used in those times were very strong. Musk and civet were the basis of most preparations, and we find them often mentioned by Shakspeare. In "Much Ado About Nothing," speaking of Benedick, Pedro says, "Nay, he rubs himself with civet : can you smell him out by that ?—that's as much as to say the sweet youth's in love." In the "Merry Wives of Windsor," Mrs. Quickly, enumerating to Falstaff all the presents made to Mrs. Ford, says, "Letter after letter, gift after gift, smelling so sweetly, all musk." With all due deference to our immortal bard, I doubt very much if a modern swain, resorting to the same means to press his suit, would find them succeed with the object of his affections ; for musk and civet used alone are anything but agreeable, and would be more likely to affect the *head* than the *heart.*

The Eastern fashion of sprinkling rose-water over the clothes seems to have been prevalent at that period ; for in one of Marston's plays a young gallant enters with a *casting bottle* of sweet water in his hand, sprinkling himself ; and in another part he says, "As sweet and neat as a barber's casting bottle."[1] Ford, in a play called "The Fairies," also mentions the same toilet implement. One of his *dramatis personæ* comes in sprinkling his hair and face with a casting bottle, and carrying a little looking-glass in his girdle, setting his countenance.

The floors of the apartments were also perfumed either with sweet rushes or with scented waters. In "Dr.

[1] Marston: Antonio and Mallida," Intr.

Faustus," an old play by Marlow, Pride enters, saying, "Fye, what a smell is here! I'll not speak another word for a king's ransom, unless the ground is perfumed.' Even in churches this used to be the case; but in summer they generally strewed flowers in the pews, instead of scents. In "Apius and Virginia," a play of that period, we find the following illustration of this habit:

> "Thou knave, but for thee ere this time of day
> My lady's fair pew had been strewed full gay
> With primroses, cowslips, and violets sweet,
> With mints, and with marygold, and marjoram meet,
> Which now lyeth uncleanly, and all along of thee."

This custom is still in vogue in Spain and Portugal, where the floor of churches is generally strewn in summer with lavender and rosemary.

Perfumes were likewise used to burn in rooms, and to fumigate sheets. "Now are the lawn sheets fumed with violets," says Marston in "What You Will." In "Much Ado About Nothing," Borachio, being asked how he came into the palace, answers, "Being entertained for a perfumer, as I was smoking a musty room," etc.; and Strype, in his "Life of Sir J. Cheke," mentions that he sent for a "perfume pan" for his apartments.[1]

Burton, in his "Anatomy of Melancholy," says, "The smoke of juniper is in great request with us to sweeten our chambers;" and in Ben Jonson we find, "He doth sacrifice twopence in juniper to her every morning before she rises, to sweeten the room by burning it."

[1] Strype's "Life of Sir J. Cheke," p. 39.   A.D. 1549.

Perfumed bellows were another device resorted to for producing a fragrant atmosphere, and Richelieu, who was a great Sybarite, made use of them in his apartments. Ford, in one of his plays, thus alludes to this custom :—

> "I'll breathe as gently
> As a perfumed pair of sucking bellows
> In some sweet lady's chamber."

Scented gloves were then usually sold by milliners or haberdashers, and various fragrant herbs were kept by apothecaries, who in London mostly dwelt in Bucklersbury, which accounts for Shakspeare's expression, "Smelling as sweet as Bucklersbury in simpling time." This fragrant herb business included all aromatics then in use, such as rosemary, which, singularly enough, was used at weddings as well as funerals, and divers woods for burning, as Beaumont and Fletcher have it in "Wit without Money :"—

> "Selling rotten wood by the pound, like spices,
> Which gentlemen often burn by the ounces."

Numerous hawkers also travelled the country, and attended country fairs, where they offered their sundry wares, like Autolycus in the "Winter's Tale :"—

> "Gloves as sweet as damask roses
> Masks for faces and for noses,
> Bugle bracelet, necklace amber,
> Perfume for a lady's chamber."

In the reign of Charles I., perfumes were extensively used as preservatives from the plague; and among the various specifics devised by the doctors of that period a curious one is mentioned by Rushworth, which consisted in eating a roasted apple stuffed with frankincense, which was recommended as a certain cure. Whether it were or not, I will not presume to say; but the prophylactic properties of scents cannot be doubted, and as late as the last century, medical practitioners carried on the top of their walking-sticks a little cassolette filled with aromatics, which they held up to their nose when they had to visit any contagious cases.

The art of facial adornment does not appear to have been far advanced at that time. We may quote as a specimen an extract from the "Poems and Fancies" of the Duchess of Newcastle, who recommends the teeth to be cleaned with "china, brick, *or the like*," and says it is customary to pull up the edges of the eyebrows by the roots, leaving none but a thin row, and to remove the first skin off the face with oil of vitriol, that a new skin may come in its place—a very strange way, certainly, of improving the complexion.

During the Commonwealth, perfumery shared the fate of all articles of luxury, and was discarded by strict Puritans; but at the restoration of Charles II., "the Merry Monarch," it was again in favour with his brilliant court. It became then customary for all ladies of fashion to paint their faces, and wear patches, which were supposed to add piquancy to the features, but which also served sometimes to conceal some disfigurement, as Pepys in his Diary represents the Duchess of Newcastle "wearing many black patches because of pimples about her mouth."[1] Some of these patches were of the most extraordinary shapes, such as suns,

[1] Pepys' Diary, 26th April, 1687.

moons, stars, etc., as Butler has it in his "Hudi-bras:"[1]—

> "The sun and moon, by her bright eyes
> Eclipsed and darken'd in the skies,
> Are but black patches that she wears,
> Cut into suns, and moons, and stars."

The annexed illustration, taken from an engraving of the period, represents a lady who, in addition to the

above, had adorned her countenance *with a coach and horses!* This custom became so prevalent, that Grammont says in his memoirs that you were always sure to find rouge and patches on a lady's toilet.

It was also the fashion then for both sexes to blacken the eyebrows, as we find in Shadwell's "Humourists"— "Be sure if your eyebrows are not black, to black 'em soundly. Ah! your black eyebrow is your fashionable eyebrow. I hate rogues that wear eyebrows that are out of fashion."

Hair-powder was introduced towards the end of the sixteenth century, probably by some person who had turned prematurely grey, and, like the fox who had lost his tail in the trap, wanted others to assume the same appearance. This fashion seems to have become extensively patronised, if we may judge from the criticisms of contemporaneous writers. Taylor, the water-poet, says in his "Superbiæ Flagellum"—

> "Some every day do powder so their hair,
> That they like ghosts or millers do appear;
> But let them powder all that e'er they can,
> Their pride will show both before God and man."

The custom of wearing hair-powder lasted about two centuries, and may be scarcely called extinct now,

Applying Hair Powder. (temp. Louis XV.)

being still patronized by eccentric *belles* and aristo-

14

cratic footmen. It certainly imparts a degree of softness to the features, but must be very inconvenient to
apply, as may be judged by the preceding engraving,
of the time of Louis XV.

The following quotation from the "Virtuoso," another
of Shadwell's plays, enumerates the various articles
which formed then the complete stock of a perfumer :
"I have choice good gloves, Amber, Orangery, Genoa,
Romane, Frangipane, Neroly, Tuberose, Jessamine, and
Marshall ; all manners of tires for the head, locks, tours,

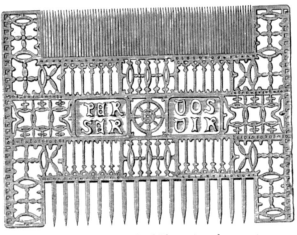

Comb of the 17th century.[1]
(*From the Sauvageot collection at the Louvre*).

frowzes, combs, and so forth ; all manner of washes,
almond water, and mercury for the complexion ; the
best pomatums of Europe, but a rare one made of lamb's

[1] This specimen forms part of a series of combs, which are all exquisitely carved. The words *Per vos Servir* (pour vous servir) engraved on it,
show it to be of foreign workmanship.

caul and May dew. Also all manner of confections of *mercury and hog's bones* to preserve present and to restore lost beauty."

The last-mentioned preparation would not appear very tempting, were it not coupled with a promise calculated to over-rule every objection : at all events, perfumers of that period must have credit for their candour in mentioning the strange ingredients which they employed.

Some historians pretend that Louis XIV., king of France, had a strong dislike for perfumes, which were consequently banished from his court. I at first shared their opinion, until, meeting accidentally with a very interesting and erudite book by M. Edouard Fournier,[1] I was convinced of my error. It appears, on the contrary, that this king was very fond of scents, and was said to be " le plus doux fleurant," or the " sweetest smelling" monarch that had ever been seen. " Le Parfumeur Françoys," a curious book published in 1680, leaves no doubt on the subject, for it says that "his Majesty was often pleased to see Mr. Martial[2] compose in his closet the odours which he wore on his sacred person." It was not then considered derogatory for great people to superintend the manufacture of their perfumes, for the Prince de Condé had his snuff scented in his presence ; and the celebrated " Poudre à la Maréchale," which still holds its place in the modern

---

[1] Paris Démoli, par Edouard Fournier.

[2] A celebrated perfumer of that period mentioned by Molière in his " Comtesse d'Escarbagnas."

perfumer's catalogue, was so named because it was at first composed by Madame la Maréchale d'Aumont.

Italy had still the privilege then of supplying the rest of Europe with the finest perfumes. When Poussin, the great French painter, went to Rome, he was entrusted by M. de Chanteloup with the mission of purchasing scented gloves, which he procured from "la Signora Maddelena," who was then in repute as the best Roman perfumer; and Du Pradel, in his "Livre Commode des adresses," mentions the "*Sieur Adam courrier de cabinet*," who often brought fine essences from Rome, Genoa, and Nice.

I have in my possession an old English book, called the "Queen's Closet," printed in 1663, which gives a complete insight into the art of perfumery at that period. It contains a number of very curious recipes, among which are those of a perfume invented by Edward VI., another composed by Queen Elizabeth, a wonderful pomatum made from apples mixed with the fat of a *young dog*, and a highly-praised dentifrice made by Mr. Ferene, of the New Exchange, perfumer to the Queen, who was, I suppose, the first of the generation. This gentleman seems to have shared the Duchess of Newcastle's partiality for bricks, for they form the chief ingredient in his tooth-powder.

Under the reign of Louis XV., perfumes still increased in favour with the French Court, and etiquette prescribed the use of a particular sort every day, which caused Versailles to be named "*la cour parfumée.*" At Choisy, also, where Madame de Pompadour held the

sceptre of elegance and beauty, perfumes were in great
favour, and formed no inconsiderable item in that
lady's household expenses, which amounted at one time
to 500,000 livres per annum.

This taste continued to prevail in France, until the
sanguinary days of the Revolution caused a momentary

Madame de Pompadour at Choisy.

interruption in the use of articles of luxury, which
returned with the advent of the imperial court. The
Empress Josephine entertained the usual passionate
fondness of creoles for scents, and her consort shared
it in no small degree.

In England, under the Georges, perfumery was more

or less in favour according to the different notions of the magnates who held by turns the sceptre of fashion.

At the commencement of the last century, the perfumer in vogue seems to have been one Charles Lilly, who lived in the Strand, at the corner of Beaufort Buildings.[1] His name is frequently mentioned in the *Tatler*, which highly praises his skill in preparing "snuffs and perfumes, which refresh the brain in those that have too much for their quiet, and gladdens it in those who have too little to know the want of it."

The next one who seems to have attracted a little notice is a Mr. Perry, residing also in the Strand, at the corner of Burleigh Street. He was, however, reduced to "blow his own trumpet;" and in a paper called the *Weekly Packet*, bearing the date of 28th December, 1718, he vaunts, besides his perfumes, an oil drawn from mustard-seed, which, at the moderate price of 6*d.* per ounce, is warranted to cure all diseases under the sun.

Some of the French perfumers of that period also combined with their " sweet wares " various sorts of medicines. This was particularly the case with the itinerant vendors or " charlatans,"[2] who, arrayed in a gorgeous red coat, with gilt lacings, addressed the gaping crowd from an elegant equipage, and dealt out their perfumes and quack remedies with musical accompaniment. The illustration forming the frontispiece to this chapter represents one of these " strolling perfumers," who

[1] By a very curious coincidence, I now occupy the same premises.
[2] From the Italian " ciarlare," to chatter.

usually sold powders, elixirs, pills, opiates, eau-de-Cologne, and scouring drops. Eight or ten years before the Revolution the King's physician had them banished from the kingdom, and from that time perfumery held a more respectable position in the industrial world. Now, thanks to the progress of science and education, it has shaken off the trammels of quackery, and become an important branch of our commerce.

I shall conclude this chapter with a few remarks on the hair and beard. The Gauls wore their hair long, whence their country derived its appellation of *Gallia Comata*, or long-haired Gaul. Julius Cæsar compelled them to cut it off when they were subdued, which they considered a great disgrace. The ancient Britons were likewise very proud of the length of their hair, of which they took great care. They shaved their chins, but preserved a long moustache. The Anglo-Saxons and Danes paid also great attention to their hair. The Danish soldiers who were quartered in England at the time of Edgar and Ethelred were the *beaux* of the period, and are said to have captivated English ladies with their fine hair, which they combed and dressed *once a day*. The clergy, who were obliged to shave the crown of their heads and keep their hair short, were constantly preaching against long hair, and even sometimes carried their precepts into action by cutting off with their own hands the hair of their flock ; but their victories were of short duration, and the favourite fashion soon resumed its sway. Men continued to wear their hair long until the time of Francis I.,

King of France, who, having been wounded in the head at a tournament, had his hair cut close, and took to wearing his beard as a compensation, which example was of course followed immediately by the whole country. This custom soon spread to England, where we

German Barber (16th century).

find it in full vigour during the reign of Henry VIII., as we may judge by Holbein's pictures, in which the head seems almost destitute of hair. The above engraving, representing a German barber in the sixteenth century, from a design by Jost Amman, illustrates this fashion, which certainly seems to facilitate the "shampooing" operation undergone by the customer at the back of the shop.

In Charles the First's time, ringlets were again in fashion for men as well as for women. "I know many young gentlemen," says Middleton in one of his plays, "wear longer hair than their mistresses." The beard was worn in various ways, the favourite shape being what Beaumont and Fletcher, in their "Queen of Corinth," denominate the T beard, consisting of the moustache and imperial:—

> " His beard,
> Which now he put i' the form of a T,
> The Roman T; your T beard is the fashion,
> And two-fold doth express the enamoured courtier."

The beard was also dyed in sundry colours, as mentioned by Shakspeare in some of his plays. The Puritans had their hair closely cropped, whence they acquired the cognomen of "Roundheads;" but long hair returned with Charles II. As, however, every one was not naturally gifted with luxuriant locks, periwigs were invented to supply the deficiency. I would rather ascribe it to this cause, for the honour of the gentlemen of the period, than to the reason given by Pepys, who says in his Diary, "At Mr. Jervas's, my old barber, I did try two or three borders and periwigs, meaning to wear one, and yet I have no stomach for it, but that *the pain of keeping my hair clean is so great.*"[1] Powder and queues came next into fashion, and were patronised during the whole of the last century, until the French

[1] Pepys' Diary, 9th May, 1663.

Revolution brought a complete change in costume and habits, and the hair was cut short, *à la Titus*, in imitation of the antique.

As regards ladies' head-dresses, the various modes adopted by turns are so numerous that it would fill a whole book to enumerate them all. The hair being the only part of a woman's charms that she can alter at her will, it has naturally been subjected to a constant change of style. In ancient times young ladies, before their marriage, used to wear their hair uncovered and untied, flowing loose over their shoulders; but when they entered the wedded state they cut it off and assumed some sort of head-gear. A little later they made it into long tresses, which sometimes reached their heels. In Richard the Second's time, the hair was worn confined in a golden net or caul—an Eastern custom, probably brought over by the Crusaders. Then came those high conical caps introduced by Isabeau de Bavière, which were made of such extraordinary dimensions that doors had to be altered to admit them. A specimen of these may still be seen in the "Pays de Caux," a part of Normandy where they are worn by rich farmers' wives. During the earlier half of the fifteenth century the

Caricature of the Horned Head-dress

horned head-dress was adopted, and its form and dimensions became the frequent butt of the satirists and caricaturists of the age. The sketch on the preceding page, from the church of Ludlow, in Shropshire, represents an aged dame whose "horns" inspire evident terror to her two companions, who appear to deem them a sign of some relationship to the spirit of evil. In Queen Elizabeth's time, flaxen hair was greatly prized as being *the queen's own colour*, and we find it frequently alluded to by poets of the period.

> "Her hair is auburn, mine is perfect yellow,"

says Julia, in the "Two Gentlemen of Verona," and Bassano, in the "Merchant of Venice," exclaims, on seeing Portia's likeness—

> "Here in her hairs
> The painter plays the spider; and hath woven
> A golden mesh to entrap the hearts of men,
> Faster than gnats in cobwebs."

False hair was also often resorted to at that time, and appears to have been varied according to the age of the wearer, if we may judge by the following epigram written by Lord Brooke :—

> "Cœlica, when she was young and sweet,
>   Adorned her hair with golden borrowed hair;
> And now in age, when outward things decay,
> In spite of age she throws the hair away,
> And now again her own black hair puts on
> To mourn for thoughts by her worth's overthrown." [1]

Under the reign of Charles II., short curls on the forehead, and ringlets at the sides, came into vogue. This was called the "Sévigné" style, and may be seen in Lely's portraits at Hampton Court Palace. In the

---

[1] Lord Brooke, p. 202.

last century head-dresses assumed the most extravagant

VARIOUS STYLES OF HEAD DRESSES, LAST CENTURY.

Butterfly.   Dove.   Battery.

Bride.   Noble Simplicity.   Great Pretensions.

Jardinière.   Tuileries   Frigate.

dimensions. They were complete edifices rising some two

Capricious.	Intercepted Looks.	Union.
Pilgrimage	Flower Girl.	Shepherdess
Porcupine	Friendship	Victory.

or three feet above the head, and comprising every

possible and impossible ornament.   The foregoing illustrations will convey some idea of these *coiffures*, the denomination of which is at least as quaint as their appearance, and which inspired the following squib, in the *London Magazine* for 1777 :—

> "Give Chloe a bushel of horsehair and wool,
>   Of paste and pomatum a pound ;
> Ten yards of gay ribbon to deck her sweet skull,
>   And gauze to encompass it round."

Of the fashions of the present century it would be needless to speak, for they are still fresh in the memory of my readers ; nor shall I presume to offer an opinion touching their respective merits.   Ladies are the best judges of what sets off their charms ; and, after all, what matters the frame when the picture is pretty ?

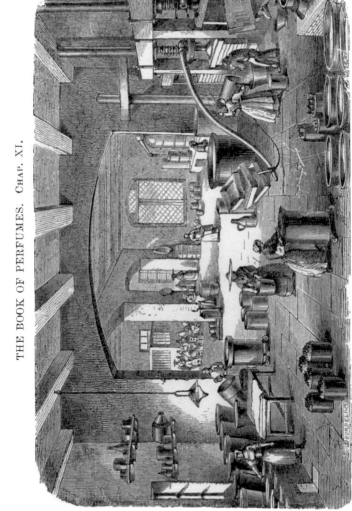

INTERIOR OF A PERFUME MANUFACTORY AT NICE.

CHAPTER XI.

THE COMMERCIAL USES OF FLOWERS AND PLANTS.

Then, were not summer's distillation left,
A liquid prisoner pent in walls of glass,
Beauty's effect with beauty were bereft,
Nor it, nor no remembrance what it was.
   But flowers distill'd, though they with winter meet,
     Leese but their show; their substance still lives sweet.

SHAKSPEARE.

NDER this heading I shall describe the various modes in use for extracting the aroma of flowers and plants. This manufacture is carried on principally in the south of France, Italy, Spain, Turkey, Algeria, India—in fact, wherever the climate gives to flowers and plants the

intensity of odour required for a profitable extraction.
The south of France furnishes the most abundant supply
of perfumery materials; there the most odoriferous
flowers—such as the rose, jasmine, orange, etc.—are cultivated on a large scale, and form the basis of the finest
perfumes. Italy produces chiefly essences of bergamot,
orange, lemon, and others of the citrine family, the
consumption of which is very great. To Turkey we
are indebted for the far-famed otto of roses, which
enters into the composition of many scents. Spain and
Algeria have yielded but little hitherto, but will no
doubt in after times turn to better account the fragrant
treasures with which nature has endowed them. Travelling in the plains of Spanish Estramadura, I have passed
through miles and miles of land covered with lavender,
rosemary, iris, and what they call "rosmarino" (*Lavandula stœchas*), all growing wild in the greatest luxuriance,
and yet they are left to "waste their sweetness on the
desert air," for want of proper labour and attention. I
also found many aromatic plants in Portugal, and among
others one named "alcrim do norte" (*Diosma ericoides*),
which has a delightful fragrance.

From British India we import cassia, cloves, sandal-
wood, patchouly, and several essential oils of the andropogon genus; and the Celestial Empire sends us the
much-abused but yet indispensable *musk*, which, carefully blended with other perfumes, gives them strength
and piquancy without being in any way offensive.

It has been proposed to cultivate flowers in England
for perfumery purposes, but the climate renders this

scheme totally impracticable. English flowers, however beautiful in form and colour they may be, do not possess the intensity of odour required for extraction, and the greater part of those used in France for perfumery would only grow here in hothouses. The only flower which could be had in abundance would be the rose, but the smell of it is very faint compared with that of the Southern rose, and the rose-water made in this country can never equal the French in strength. If we add to this the shortness of the flowering season, and the high price of land and labour, we may arrive at the conclusion that such a speculation would be as bad as that of attempting to make wine from English grapes. As a proof of this, I may mention that I had a specimen submitted to me not long since of a perfumed pomade which a lady had attempted to make on a *flower-farm* which she had been induced to establish in the north of England, and it was, as I expected, a complete failure.

The only two perfumery ingredients in which England really excels are lavender and peppermint, but that is owing to the very cause which would militate against the success of other flowers in this country; for our moist and moderate climate gives those two plants the mildness of fragrance for which they are prized, whilst in France and other warm countries they grow strong and rank.

There are four processes in use for extracting the aroma from fragrant substances—distillation, expression, maceration, and absorption.

Distillation is employed for plants, barks, woods, and a few flowers. These are placed in a still containing water, which evaporates by means of heat, condenses in the worm, and issues from the tap strongly impregnated with the aroma, the more concentrated part of which collects either on the surface or at the bottom of the distillate, according to its specific gravity, and forms the essential oil. The same water is generally distilled

Steam Still.

several times over with fresh materials, and is sometimes of sufficient value to be kept, as is the case with rose and orange-flower water. A great improvement has been lately introduced in the mode of distillation : it consists in suspending the flowers or plants in the still on a sort of sieve, and allowing a jet of steam to pass through and carry off the fragrant molecules. This produces a finer essential oil than allowing these substances to be steeped in water at the bottom of the still.

Expression is confined to the essences obtained from the rinds of the fruits of the citrine series, comprising lemon, orange, bigarrade, bergamot, cedrat, and limette. It is performed in various ways: on the coast of Genoa they rub the fruit against a grated funnel; in Sicily they press the rind in cloth bags; and in Calabria, where the largest quantity is manufactured, they roll the fruit between two bowls, one placed inside the other, the concave part of the lower and the convex part of the upper being armed with sharp spikes. These bowls revolve in a contrary direction, causing the small vesicles on the surface of the fruit to burst and give up the essence contained in them, which is afterwards collected with a sponge. These rinds are also sometimes distilled; but the former process, which is called in French *au zest*, gives a much purer essence.

Maceration and *absorption* are both founded on the affinity which fragrant molecules have for fatty bodies, becoming more readily fixed into them than into any others. Thus the aroma of flowers is first transferred to greases (called pomades), and oils, which are made afterwards to yield it to alcohol, whilst the latter, if placed in direct contact with the flowers, would not extract it from them. The first attempt that was made in this way, some two hundred years ago, was to place some almonds in alternate beds with fresh-gathered flowers, renewing the latter several days, and afterwards pounding the almonds in a mortar, and pressing the oil which had absorbed the aroma. This is the same process now used in India by the natives for ob-

taining perfumed oils, substituting gingelly or sesamum
seeds for almonds. The next improvement was to use
a plain earthen pan, coated inside with a thin layer of
grease, strewing the flowers on the grease, and covering
it over with another jar similarly prepared. After renewing the flowers for a few days, the grease was found
to have borrowed their scent. This process was aban-
doned in France some fifty years ago, but is still
resorted to by the Arabs (who were probably the
inventors of it), the only difference being that they
use white wax mixed with grease, on account of the
heat of the climate.

The two modes now adopted to make these scented
oils and pomades are, as I said before, *maceration* and
absorption. The former is used for the less delicate
flowers, such as the rose, orange, jonquil, violet, and
cassie (*Acacia farnesiana*). A certain quantity of grease
is placed in a pan fitted with a water bath, and is
brought to an oily consistency. Flowers are then
thrown in, and left to digest for some hours, being
stirred frequently; after which, the grease is taken out
and pressed in horsehair bags. This operation is re-
peated, until the fatty body is sufficiently impregnated
with the fragrance of the flowers. Oil is treated in the
same way, but requires less heat.

The process of *absorption*, called by the French *enfleurage*, is chiefly confined to the jasmine and tuberose,
the delicate aroma of which would be injured by
heat. A series of square glass frames are covered
with a thin layer of purified grease, in which ridges

are made to facilitate absorption. Fresh-gathered flowers are strewed on these, and renewed every morning as long as the flower is in bloom, and by that time the grease has acquired a very strong flavour. The same process is used for oil, but the frames, instead of being mounted with glass, have a wire bottom, over which is spread a thick cotton cloth soaked in olive oil. Flowers are laid on in the same way, and the cloths submitted to high pressure to extract the oil when suf-

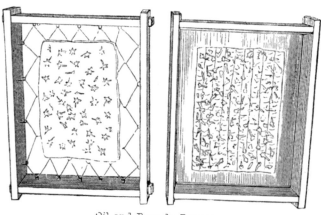

Oil and Pomade Frames.

ficiently impregnated These frames are piled on each other to keep them air-tight.

A new mode of *enfleurage* has been lately devised by Mr. D. Séméria, of Nice, and found to offer advantages over that just described. Instead of laying the flowers on the grease, he spreads them on a fine net mounted on a separate frame. This net is introduced between two glass frames covered on both sides with grease. The whole series of frames is inclosed in an air-tight recess, and all that is required is to draw out the nets every morning and fill them with fresh flowers, which give their aroma to the two surfaces with which they are in contact. This system saves the waste and labour resulting from having to pick the old flowers from the surface of the grease, and produces also a finer fragrance.

A very curious pneumatic apparatus for the same purpose has been invented by Mr. Piver, the eminent

View of Grasse.

Parisian perfumer, who submitted a plan of it to the jury at the last Exhibition. It consists in a series of perforated plates, supporting flowers placed alternately with sheets of glass overlaid with grease, in a chamber through which a current of air is made to pass several times, until all the scent of the flowers becomes fixed into the grease.

A no less remarkable invention is that of Mr. Millon, a French chemist, who found means to extract the aroma of flowers by placing them in a percolating apparatus and pouring over them some ether, or sulphuret of carbon, which is drawn off a few minutes after, and carries with it all the fragrant molecules. It is afterwards dis-

View of Nice.

tilled to dryness, and the result obtained is a solid waxy mass possessing the scent of the flower in its purest and most concentrated form. This process, although very ingenious, has not received any practical application as yet, owing to the expense attending it, some of these concrete essences costing as much as £50 an ounce. It has, however, served to prove the total imponderability of fragrant molecules ; for although this substance, from its high state of concentration, appears at first sight to be the solidified principle of scent, if it be treated several times with alcohol it gradually loses all its perfume, and yet the residue is not found to have lost one atom of its weight.

Grasse, Cannes, and Nice, all in the south of France, and close to each other, are the principal towns where the maceration and absorption processes are carried on. There are above one hundred houses engaged in these operations, and in the distillation of essential oils, giving employment during the flower season to at least ten thousand people. Nice is, perhaps, the most admirably situated of the three for producing all flowers for perfumery purposes, and its violets in particular are superior to any other. Since that town has become French a great impulse has been given to its manufacture of perfumery materials, which had formerly to pay customs duties on entering into France.

The following are approximate quantities and values of the flowers consumed in that locality for preparing perfumery materials :—

Orange-flowers	2,000,000 lbs.	worth about	£40,000
Roses	600,000 ,,	,,	12,000
Jasmine	150,000 ,,	,,	8,000
Violets	60,000 ,,	,,	4,000
Cassia	80,000 ,,	,,	6,000
Tuberose	40,000 ,,	,,	3,000

These flowers are procured from growers by private contract or sold in the market. The average quantities of the following articles are manufactured with them yearly:—700,000 lbs. of scented oils and pomades, 200,000 lbs. of rose-water, 1,200,000 lbs. of orange-flower water, first quality,[1] 2,400,000 lbs. of orange-flower water, second quality; 1,000 lbs. of neroly, an essential oil obtained from orange-flowers. The other flowers do not yield essential oils, but the latter are extensively distilled in the same places from aromatic plants, such as lavender, rosemary, thyme, geranium, etc. Many of my fair readers have considered flowers hitherto as simply ornamental: the above figures will give them an idea of their importance as an article of commerce.

Another branch of the art of perfumery is the manufacture of scents, cosmetics, soaps, and other toilet requisites. It is carried on in the principal cities of Europe, and especially in London and Paris, which may be called the head-quarters of perfumery, and whence these products are exported to all parts of the world. There are, it is true, other manufactories in Germany, Russia, Spain, and the United States, but their chief trade consists in counterfeiting the articles of the London and Paris manufacturers, and this cannot be considered a legitimate business.

The principal English manufacturers of perfumery and toilet soaps reside in London, where they number about sixty, employing a large number of men and women; for female labour has been introduced for nearly twenty years in all the London manufactories,[2] and

---

[1] That is, distilled twice over the flowers.

[2] I believe I was the first to employ female labour in *England*, and I am happy to say my example was soon followed by my *confrères*.

has been found to answer very well for all kinds of work requiring more dexterity than strength.

According to the official returns published, the exports of perfumery from the United Kingdom for the year 1863, amounted to £106,989, sub-divided as will be seen in the following table; we must, however, say that very little reliance is to be placed on these figures, which do not represent perhaps one-fourth of the actual amount exported. Taking, for instance, the sum given for Australia at £18,921, it appears ridiculously small; there are undoubtedly several manufacturers in London who each and individually ship perfumery to nearly that amount every year, to our Australian colonies.

EXPORTS OF PERFUMERY FROM THE UNITED KINGDOM IN 1863.

Countries to which Exported.	Amounts declared.
Russia	£ 2732
Hamburg	3118
Holland	1980
Belgium	2568
France	2250
Egypt	1968
China	5749
United States	4477
Brazil	2149
British Possessions in South Africa	1818
Mauritius	2141
British India	21914
Australia	18921
British North America	3415
British West Indies	6004
Channel Islands	10189
Gibraltar	1003
Portugal, Azores, and Madeira	1172
Spain and Canaries	2021
Argentine Confederation	1717
Other Countries	9683
	£ 106,989

This table does not include soap; but as perfumed soaps are not particularized, and are confounded with common ones, it is impossible to obtain any correct information respecting the amount or quantity exported.

Paris is the great centre of the manufacture of perfumery, which forms an important item of what are called "articles de Paris." There are in that capital one hundred and twenty working perfumers, employing about three thousand men and women, and their united returns may be estimated at not less than forty millions of francs yearly. The amount of perfumery exported from France alone reaches annually upwards of thirty millions of francs, its principal consumers being Europe and North and South America; whilst British perfumery is more frequently shipped to India, China, and Australia.

Next to Hungary-water, the most ancient perfume now in use is eau-de-Cologne, or Cologne-water, which was invented in the last century by an apothecary residing in that city. It can, however, be made just as well anywhere else, as all the ingredients entering into its composition come from the South of France and Italy. Its perfume is extracted principally from the flowers, leaves, and rind of the fruit of the bitter orange, and other trees of the *Citrus* species, which blend well together, and form an harmonious compound.

Toilet vinegar is a sort of improvement on eau-de-Cologne, containing balsams and vinegar in addition. Lavender-water was formerly distilled with alcohol from

fresh flowers, but is now prepared by simply digesting
the essential oil in spirits, which produces the same
result at a much less cost. The finest is made with
English oil, and the common with French, which is
considerably cheaper, but is easily distinguished by its
coarse flavour.

Perfumes for the handkerchief are composed in various ways: the best are made by infusing in alcohol
the pomades or oils obtained by the processes I have
just described. This alcoholate possesses the true scent
of the flowers entirely free from the empyreumatic smell
inherent in all essential oils; as, however, there are but
six or seven flowers which yield pomades and oils, the
perfumer has to combine these together to imitate all
other flowers. This may be called the truly artistic
part of perfumery, for it is done by studying resem-
blances and affinities, and blending the shades of scent
as a painter does the colours on his palette. Thus, for
instance, no perfume is extracted from the heliotrope;
but as it has a strong vanilla flavour, by using the latter
as a basis, with other ingredients to give it freshness, a
perfect imitation is produced; and so on with many
others.

The most important branch of the perfumer's art is
the manufacture of toilet soaps. They are generally
prepared from the best tallow soaps, which are remelted,
purified, and scented. They can also be made by what
is called the cold process, which consists in combining
grease with a fixed dose of lees. It offers a certain
advantage to perfumers for producing a delicately-

scented soap, by enabling them to use as a basis a pomade instead of fat, which could not be done with the other process, as the heat would destroy the fragrance. This soap, however, requires being kept for some time before it is used, in order that the saponification may become complete. Soft soap, known as shaving cream, is obtained by substituting potash for soda lees, and transparent soap by combining soda soap with alcohol. Another sort of transparent soap has been produced lately by incorporating glycerine into it, in the proportion of about one-third to two-thirds of soap.

The English toilet soaps are the very best that are made : the French come next, but, as they are not re-melted, they never acquire the softness of ours. The German soaps are the very worst that are manufactured : the cocoa-nut oil, which invariably forms their basis, leaves a strong fœtid smell on the hands, and their very cheapness is a deception; for as cocoa-nut oil takes up twice as much alkali as any other fatty substance, the soap produced with it wastes away in a very short time.

Cosmetics, pomatums, washes, dentifrices, and other toilet requisites, are also largely manufactured, but they are too numerous to be described here at full length ; nor shall I attempt to descant on their respective merits, which depend, in a great measure, upon the skill of the operator, and the fitness and purity of the materials used. The greatest improvement effected in these preparations lately has been the introduction of glycerine. Although this substance was discovered in the last century, it is only a few years since medical men fully recognised and appreciated its merits, and applied it to the cure of skin diseases, for which it answers admirably. Perfumers are now beginning to avail themselves of its wonderful properties, and to combine it with their soaps and cosmetics.

The volatilisation of perfumes by means of steam is also a modern improvement. A current of steam is made to pass through a concentrated essence, from which it disengages the fragrant molecules, and spreads them through the atmosphere with extraordinary rapidity and force. A whole theatre may be perfumed by this means in ten minutes, and a drawing-room consequently in much less time. This system has the advantage of purifying the air, and has been adopted on that account by some of the hospitals and other public institutions.

Before concluding this chapter I shall venture to offer to ladies a few words of advice on the choice of their perfumes and cosmetics. I feel that this is delicate ground, but I shall endeavour to let my remarks be of a purely general character.

The selection of a perfume is entirely a matter of taste, and I should no more presume to dictate to a lady which scent she should choose, than I would to an epicure what wine he is to drink ; yet I may say to the nervous : use simple extracts of flowers which can never hurt you, in preference to compounds, which generally contain musk and other ingredients likely to affect the head. Above all, avoid strong, coarse perfumes ; and remember, that if a woman's temper may be told from her handwriting, her good taste and good breeding may as easily be ascertained by the perfume she uses. Whilst a *lady* charms us with the delicate ethereal fragrance she sheds around her, aspiring vulgarity will as surely betray itself by a *mouchoir* redolent of common perfumes.

Hair preparations are like medicines, and must be varied according to the consumer. For some pomatum is preferable, for others oil, whilst some, again, require neither, and should use hair-washes or lotions. A mixture of lime-juice and glycerine has lately been introduced, and has met with great success, for it clears the hair from pellicles, the usual cause of premature baldness. For all these things, however, personal experience is the best guide.

Soap is an article of large consumption, and some people cannot afford to pay much for it; yet I would say, avoid *very cheap* soaps, which irritate the skin owing to the excess of alkali which they contain. Good soaps are now manufactured at a very moderate price by the principal London perfumers, and ought to satisfy the most economical. White, yellow, and brown are the best colours to select.

Tooth-powders are preferable to tooth-pastes. The latter may be pleasanter to use, but the former are certainly more beneficial.

Lotions for the complexion require of all other cosmetics to be carefully prepared. Some are composed with mineral poisons, which render them dangerous to use, although they may be effectual in curing certain skin diseases.

There ought to be always a distinction made between those that are intended for healthy skins, and those that are to be used for cutaneous imperfections; besides, the latter may be easily removed without having recourse to any violent remedies.

Paints for the face I cannot conscientiously recommend. Rouge is innocuous in itself, being made of cochineal and safflower; but whites are often made of deadly poisons, such as cost poor Zelger his life a few months since.[1] The best white ought to be made of mother-of-pearl, but it is not often so prepared. To professional people, who cannot dispense with these, I must only recommend great care in their selection; but to others I would say, cold water, fresh air, and exercise, are the best recipes for health and beauty; for no borrowed charms can equal those of

"A woman's face, with Nature's own hand painted."

[1] M. Zelger was a Belgian singer at the Royal Italian Opera. During the performance of "Guillaume Tell," some of the paint which he had on his face accidentally entered his mouth, and he died in consequence, after a very painful and lingering illness.

A FLOWER GARDEN AND DISTILLERY AT NICE.

CHAPTER XII.

MATERIALS USED IN PERFUMERY.

Cheiro suave, ardente especiaria.
CAMOENS.

AVING now concluded the history of perfumery, both ancient and modern, it remains for me to give a brief description of the various materials used for that branch of manufacture, which are supplied by all parts of the world, from the parched regions of the torrid zone to the icy realms of the Arctic pole.

They may be divided, according to their nature, into twelve series, viz., the animal, floral, herbal, andropogon, citrine, spicy, ligneous, radical, seminal, balmy or resinous, fruity and artificial.

The animal series comprises only three substances— musk, civet, and ambergris. It is very useful in perfumery, on account of its powerful and durable aroma, which resists evaporation longer than any other.

Musk is a secretion found in a pocket, or pod, under the belly of the musk-deer (*Moschus moschatus* or *moschiferus*), a ruminant which inhabits the higher moun-

Musk-Deer (*Moschus moschiferus.*)

tain ranges of China, Thibet, and Tonquin. "It is a pretty grey animal," says Dr. Hooker, "the size of a roebuck, and somewhat resembling it, with coarse fur, short horns, and two projecting teeth from the upper jaw, said to be used in rooting up the aromatic herbs from which the Bhoteas believe that it derives its odour."[1] The male alone yields the celebrated perfume, the best being that which comes from Tonquin. The next in quality is collected in Assam; whilst the Kaberdeen musk, obtained from a variety of the species

[1] Himalayan Journal, by Dr. Hooker, vol. i., p. 256.

called Kubaya (*Moschus Sibiricus*), which inhabits the Siberian side of those mountains, is the most inferior of all.

The Chinese have known musk for many ages: they call it *shay hëang,*—*shay* being the name of the animal, and *hëang* meaning perfume. Tavernier is the first European traveller who mentions the precious drug, and he says he bought 7673 pods in one of his journeys, which shows how plentiful it must have been even at that early period. He gives the following description of musk-deer hunting, which takes place in February and March, when hunger drives these animals from

Musk-Deer Hunting. (*From a Chinese drawing.*)

their wild snowy haunts towards cultivated regions:—
'At that time the hunters lie in wait for them with

snares, and kill them with arrows and sticks. They are so lean and exhausted through the hunger they have endured, that they are easily pursued and over-taken."[1] The foregoing illustration, faithfully copied from a Chinese drawing, in which were wrapped up some musk-pods I purchased lately, would tend to prove that the musk-deer chase is still carried on in the same manner.

Musk is an unctuous substance of a reddish-brown colour, which soon becomes black by exposure to the air.

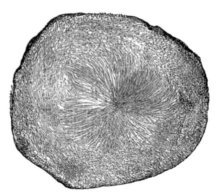

Musk Pod. (Natural size)

It is so powerful that, according to Chardin's authority, the hunter is obliged to have his mouth and nose stopped with folds of linen when he cuts off the bag from the animal, as otherwise the pungent smell would cause hæmorrhage, sometimes ending in death. As, however, the natives take good care to adulterate the musk before they send it to Europe, we are not exposed to such accidents. The substances used for this adul-

[1] Voyage de Jean Baptiste Tavernier, vol. iv., p. 75.

teration are generally the blood or chopped liver of the animal, which they cleverly insert into the pod, and sometimes pieces of lead are introduced to increase the weight. Some even manufacture artificial pods from the belly skin, and fill them with a mixture of musk and other materials. Musk, in pods, is generally imported in caddies of twenty ounces in weight, and the price of it varies from 25s. to 50s. per ounce, according to quality. *Grain musk*, which is the musk extracted from the pods, is much dearer. Musk is, without any exception, the *strongest* and *most durable* of all known perfumes, and it is, in consequence, largely used in compounds, its presence, when not too perceptible, producing a very agreeable effect.

The odour of musk is not confined to this species of animals: it is also to be found, though in a less degree, in others, such as the musk-ox, the musk-rat, the musk-duck, etc. Chief Justice Temple, of British Honduras, who presided at the Society of Arts when I read my paper on perfumery, assured the meeting that the glands of alligators had a strong musky odour; and, wishing to ascertain the fact, I procured, through the kindness of my friend, Mr. Edward Greey, of the Royal West India Mail Company, the head of one of these monsters: but I must say that, when the case was opened, the stench diffused was so great that it required some little amount of courage to extract the glands, and the *perfume* they seemed to possess was strongly suggestive of Billingsgate market on a hot day. Some polypi, and, among others, the *Tipula moschifera*, which is found in the Mediterranean, and principally at Nice, give out a musky smell, but of a very evanescent nature.

The musky fragrance likewise occurs in some vegetables, such as the well-known yellow-flowered musk-plant, but its intensity is not sufficient for extraction. The definition *moschatus* (musky), is often applied to plants and flowers; but it must not always be taken in its literal sense, for botanists are apt to distinguish by this name strong scents, such as the nutmeg, which is termed *Myristica moschata*, although it bears no resemblance to musk. The so-called *musk-seed*, itself (*Hibiscus abelmoschus*) is much more like civet than musk. Dr. Cloquet pretends that some preparations of gold and other mineral substances have also a musky fragrance,[1] but I have never met with any which bore out this assertion.

Civet is the glandular secretion of the *Viverra civetta*,

Civet Cat (*Viverra civetta*)

an animal of the feline tribe, about three feet in length and one foot in height, which is found in Africa and

[1] Osphrésiologie, p. 76.

India. It is now chiefly imported from the Indian Archipelago; but, formerly, Dutch merchants kept some of these cats at Amsterdam in long wooden cages, and had the perfume scraped from them two or three times a week with a wooden spatula. Civet, in the natural state, has a most disgusting appearance, and its smell is equally repulsive to the uninitiated, who would be tempted to cry out with Cowper—

> "I cannot talk with civet in the room,
> A fine puss gentleman that's all perfume;
> The sight's enough, no need to smell a beau
> Who thrusts his nose into a raree show."

Yet, when properly diluted and combined with other scents, it produces a very pleasing effect, and possesses a much more *floral* fragrance than musk; indeed, it would be impossible to imitate some flowers without it. Its price varies from 20s. to 30s. per ounce, according to quality.

Ambergris for a long time puzzled the *savans*, who were at a loss to account for its origin, and thought it at first to be of the same nature as yellow amber, whence it derived its name of *grey amber* (*ambre gris*). It is now ascertained beyond a doubt to be generated by the large-headed spermaceti whale (*Physeter macrocephalus*), and is the result of a diseased state of the animal, which either throws up the morbific substance, or dies of the malady, and is eaten up by other fishes. In either case, the ambergris becomes loose, and is picked up floating on the sea, or is washed ashore. It is found principally on the coasts of Greenland, Brazil, India, China, Japan, etc., and sometimes on the west coast of

Ireland. The largest piece on record was one weighing
182 lbs., which the Dutch East India Company bought
from the King of Tydore. I have in my possession a
very curious specimen extracted by a North American
whaler from a fish which he killed. Part of it is quite
grey, and the remainder still black, which shows that
the disease had not yet attained its maturity.

Ambergris is not agreeable by itself, having a somewhat earthy or mouldy flavour, but blended with other
perfumes it imparts to them an ethereal fragrance un-
attainable by any other means. Its price varies very
much, according to the quantity to be found in the
market. I have known it as low as 10s. and as high as
50s. per ounce.

The floral series includes all flowers available for
perfumery purposes, which hitherto have been limited
to eight—viz., jasmine, rose, orange, tuberose, cassie,
violet, jonquil, and narcissus.

Jasmine is one of the most agreeable and useful
odours employed by perfumers, and highly valuable are
the fragrant treasures which they obtain

> " From timid jasmine buds, that keep
>   Their odours to themselves all day,
>   But, when the sunlight dies away,
>   Let their delicious secret out." [1]

It was introduced by the Arabs, who called it Yasmyn,
hence its present name. The most fragrant sort is the
Jasminum odoratissimum, which is largely cultivated
in the south of France. It is obtained by grafting on
wild jasmine, and begins to bear flowers the second

[1] Light of the Harem.

year. It grows in the shape of a bush from three to four feet high, and requires to be in a fresh open soil, well sheltered from north winds. The flowering season is from July to October. The flowers open every morning at six o'clock with great regularity, and are culled after sunrise, as the morning dew would injure their flavour. Each tree yields about twenty-four ounces of flowers.

We next come to the queen of flowers, the rose—the eternal theme of poets of all ages and of all nations, but which for the prosaical perfumer derives its principal charms from the delicious fragrance with which Nature has endowed it.

> " The rose looks fair, but fairer we it deem
> For that sweet odour which doth in it live." [1]

And well does the perfumer turn that sweetness to account ; for he compels the lovely flower to yield its aroma to him in every shape, and he obtains from it an essential oil, a distilled water, a perfumed oil, and a pomade. Even its withered leaves are rendered available to form the ground of sachet-powder, for they retain their scent for a considerable time.

The species used for perfumery is the hundred-leaved rose (*Rosa centifolia*). It is extensively cultivated in Turkey, near Adrianople, whence comes the far-famed otto of roses ; and in the south of France, where pomades and oils are made.

Rose trees are planted in a cool ground, and may be exposed to the north wind without any injury. They

---

[1] Shakspeare's Sonnets, liv.

bear about eight ounces of flowers in the second year, and twelve ounces in the following ones. The flowering season is in May, and the flowers, which generally open during the night, must be gathered before sunrise, as after that time they lose half their fragrance.

The orange-blossoms used for perfumery are those of the bigarrade or bitter orange-tree (*Citrus bigarradia*). They yield by distillation an essential oil known under the name of *néroly*, which forms one of the chief ingredients in eau-de-Cologne : a pomade and an oil are also obtained from them by maceration. From the leaves of the tree an essential oil called *petit-grain* is produced, and from the rind of the fruit another essence is expressed, which is styled *oil of bigarrade*. The edible orange-tree (*Citrus aurantium*) also produces essences, but they are of a very inferior quality, with the exception of that obtained from the rind, which is called *oil of Portugal*. These two trees bear a great resemblance to each other, but the petiole of their leaves are slightly different; that of the *bigarrade* being in the shape of a heart.

The largest bigarrade-tree plantations are to be found in the south of France, in Calabria, and in Sicily. This tree requires a dry soil, with a southern aspect. It bears flowers three years after grafting, increasing every year until it reaches its maximum, when it is about twenty years old. The quantity depends on the age and situation, a full-grown tree yielding on an average from 50 lbs. to 60 lbs. of blossoms. The

Bigarrade
Leaf.

flowering season is in May, and the flowers are gathered two or three times a week, after sunrise.

The tuberose (*Polyanthes tuberosa*) is a native of the East Indies, where it grows wild, in Java and Ceylon : it was first brought to Europe by Simon de Tovar, a Spanish physician, in 1594. The Dutch monopolised this flower for some time, cultivating it in hothouses, but it has now found its way to France, Italy, and Spain, and thrives well in those climates.

> " Eternal spring, with smiling verdure here
>   Warms the mild air, and crowns the youthful year.
>   The tuberose ever breathes, and violets blow."

It springs from a bulb which is planted in the autumn and bears flowers the following year. The stalk rises about three feet, and produces every day two full-blown flowers, which open from 11 A.M. to 3 P.M., according to localities, but always with the most precise regularity : they

Tuberose (*Polyanthes tuberosa*)

must be gathered immediately, as their fragrance does not last long.

Cassie (*Acacia farnesiana*) is a shrub of the acacia tribe, which only grows in southern latitudes. Its height ranges from five to six feet, and it becomes covered in the months of October and November with globular flowers of a bright golden hue, which, peering through its delicate emerald foliage, have the prettiest effect. All those who have travelled in that season on the coast of Genoa will no doubt remember what charming bouquets and garlands are made of the cassie intermixed with other flowers. To perfumers it is a most valuable assistant, possessing in the highest degree

Cassie (*Acacia farnesiana*).

a fresh floral fragrance, which renders it highly useful in compounds. It bears some resemblance to the violet, and, being much stronger, is often used to fortify that scent, which is naturally weak.

The cassie requires a very dry soil, well exposed to the sun's rays. The tree does not bear flowers until it is five or six years old. The yield varies from 1 lb. to 20 lbs. for every tree, according to age and position. The blossoms are gathered three times a week after sunrise: a very strong oil and pomade is obtained from them by maceration. In Africa, and principally in Tunis, an essential oil of cassie is made, which is sold at about £4 per ounce; but French and Italian flowers are not sufficiently powerful to yield an essence.

The violet is one of the most charming odours in nature, and well might Shakspeare exclaim—

"Sweet thief, whence didst thou steal thy sweet that smells,
    If not from my love's breath?"

It is a scent which pleases all, even the most delicate and nervous, and it is no wonder that it should be in such universal request. The largest and almost only

violet plantations have hitherto been at Nice, its excep-
tional position rendering it the most available spot for
them. The species used is the double Parma violet
(*Viola odorata*). It requires a very cool and shady
ground, and is generally placed in the orange and
citron groves, at the foot of the trees, which screen it
with their thick foliage from the heat of the sun. It
flowers from the beginning of February to the middle
of April, and each plant yields but a few ounces of
blossoms, which are culled twice a week after sunrise.

Jonquil (*Narcissus jonquila*), and narcissus (*Narcissus
odorata*), are two bulbous plants which are also cultivated for perfumery purposes, but in much smaller
quantities than any of those already mentioned, their
peculiar aroma rendering their use limited. The former
is to be found chiefly in the south of France, and the
latter in Algeria. Mignonnette, lilac, and hawthorn
are also sometimes worked into pomades, but on such
a small scale that they are not worth mentioning. The
extracts named after those flowers are generally produced by combination.

The herbal series comprises all aromatic plants, such
as lavender, spike, peppermint, rosemary, thyme, marjoram, geranium, patchouly, and wintergreen, which
yield essential oils by distillation.

Lavender was extensively used by the Romans in
their baths, whence it derived its name.[1] It is a nice,
clean scent, and an old and deserving favourite. The
best lavender (*Lavandula rera*) is grown at Mitcham,

[1] From the Latin *lavare*, "to wash."

in Surrey, and at Hitchin, in Hertfordshire. It is produced by slips, which are planted in the autumn, and
yield flowers the next year and the two following ones,
when they are renewed. Mr. James Bridges, the largest
English distiller of lavender and peppermint, cultivates
these two plants on an extensive scale near Mitcham.
During the flower season he has three gigantic stills in
operation, each able to contain about one thousand gallons.

A great deal of essence of lavender is also manufactured in France; but, as I said before, it is very inferior
to that made in England. It is obtained from the
same plant, which grows wild in great abundance in
most Alpine districts. Portable stills are carried into
the mountains, and the herb distilled on the spot.
The same process is used for rosemary and thyme.

Spike (*Lavandula spica*) is a coarser species of lavender, which is principally used for mixing with the
other, or for scenting common soaps. A third sort of
lavender (*Lavandula stœchas*) has a beautiful odour, and
would yield a very fragrant essence, but it is very
scarce in France: the only places where I met with it
in quantities are Spain and Portugal, and there it is
only used to strew the floors of churches and houses on
festive occasions, or to make bonfires on St. John's day, a
custom formerly observed in England with native plants.

Peppermint (*Mentha piperita*) is more used by con-
fectioners than perfumers, yet the latter find it useful
in tooth-powders and washes. It is, like lavender, best
grown in England, the foreign being very inferior.
The American comes next to the English in quality.

Rosemary (*Rosmarinus officinalis*) is another plant of the labiate order, which yields a powerful essence, used chiefly for scenting soap. The resemblance of its flavour to that of camphor is very remarkable.

There are two sorts of thyme distilled—ordinary thyme (*Thymus vulgaris*), and wild thyme, or serpolet (*Thymus serpyllum*). Marjoram (*Origana majorana*) belongs to the same class.

The rose-geranium (*Pelargonium odoratissimum*) yields an essence which is greatly prized by perfumers on account of its powerful aroma, by means of which they impart a *rosy fragrance* to common articles at a much less cost than by using otto of roses, which is worth six times as much. It is cultivated in the south of France, Algeria, and Spain. The latter produces the finest essence, which is principally obtained from the fertile " Huerta de Valentia."

Patchouli (*Pogostemon patchouli*) comes from India, where it is known under the name of *puchaput*. It has a most peculiar flavour, which is as offensive to some as it is agreeable to others.

Patchouli.

Wintergreen (*Gaultheria procumbens*) we receive from North America. This essence is exceedingly powerful, and requires to be used with great caution to produce a pleasing effect. Well blended with others in soap, it imparts to it a rich *floral* fragrance.

The andropogon[1] series embraces three sorts of aromatic grasses, which grow abundantly in India, and principally in Ceylon, whence we obtain their essential oils. They are the *Andropogon schœnanthus*, or lemongrass, which is used to imitate verbena, having a somewhat similar fragrance; the *Andropogon citratum*, or citronella, which forms the basis of the perfume of honey soap; and the *Andropogon nardus*, or gingergrass oil, improperly called Indian geranium, which I have already mentioned in Chapter VIII. The chief use of the latter in the East, I am sorry to say, is to adulterate otto of roses, which costs from 30s. to 40s. per ounce, whilst the other oil is scarcely worth one shilling per ounce.

The citrine series comprises bergamot (*Citrus bergamia*), sweet orange (*Citrus aurantium*), bitter orange (*Citrus bigarradia*), lemon (*Citrus medica*), cedrat (*Citrus cedrata*), and limette (*Citrus limetta*). Essential oils are expressed or distilled from the rind of all these fruits, as described in the last chapter.

The spice series includes cassia, cinnamon, cloves, mace, nutmeg, and pimento.

Cassia, which was, like cinnamon, well known and highly prized by the ancients, is distilled from the *Laurus cassia*, a tree of the laurel tribe, which is abundant in the East Indies and China.

Cinnamon belongs to the same class, and is extracted from the bark of the *Laurus cinnamomum*. A coarser

---

[1] From ανδρος πώγον, so called because this grass resembles a man's beard.

essence is likewise obtained from the leaves of the same tree.

Cloves are the flower-buds of the *Caryophyllus aromaticus*, a tree found in the Indian Archipelago. The finest come from Zanzibar. The essence is chiefly used for scenting soap; but, when in infinitesimal quantities, it also blends well with some handkerchief scents, and principally with the carnation and clove-pink, the fragrance of which it closely resembles.

Cloves   Nutmeg.

Mace and nutmeg are both produced by the *Myristica moschata*, the latter being the fruit of that tree, and the former one of its envelopes, or husks.

Pimento, or allspice, is the berry of the *Eugenia pimenta*, from which an essential oil is distilled, which, like the two last named, is used for perfuming soap.

The ligneous series consists of sandal-wood, rose-wood, rhodium, cedar-wood, and sassafras.

Sandal-wood comes from the East, where it is highly

esteemed as *the* perfume *par excellence*, forming the ground of all toilet preparations. There are several species, the best being the *Santalum citrinum*, from which the essential oil used by perfumers is chiefly distilled. I observed, in the last Exhibition, some very fine specimens from Western Australia and New Caledonia.

Rosewood (*Lignum aspalathum*), rhodium (*Convolvulus scoparia*), and cedar-wood (*Juniperus virginiana*) likewise yield essential oils, which are, however, but little used by perfumers.

Benzoin.   Camphor.

Sassafras, distilled from the *Laurus sassafras*, a tree which grows abundantly in North America, is a very useful essence for soap, on account of its fresh and powerful aroma.

The radical series is confined to orris-root and vetivert.

Orris, or iris, is the rhizome of the *Iris Florentina*, which is extensively cultivated in Italy, and principally in Tuscany. It exhales, when dry, a delightful violet fragrance, which renders it very useful for scenting toilet, sachet, and tooth powders. When infused in spirits it loses the violet odour, owing to the resinous matters contained in it, which become dissolved and overpower it; but it is still sufficiently pleasant to form the basis of many cheap perfumes.

Vetivert, or kus-kus, is the rhizome of the *Anatherum muricatum*, which grows wild in India, as mentioned in a former chapter. It forms the basis of the perfume called *mousseline*, which derived its name from the

Dipterix Odorata

Sassafras.

peculiar odour of *Indian muslin*, which had formerly great repute in Europe, and which was scented with this root by the natives. Some of the Cyprus species in India also possess fragrant roots, but they are little used in Europe.

The seminal series includes aniseed (*Pimpinella anisum*), dill (*Anethum graveolens*), fennel (*Anethum fœniculum*), and carraway (*Carum carui*), all umbelliferous plants, with aromatic seeds which yield essential oils. The last-named is the most largely used. Musk-seed, obtained from the *Hibiscus abelmoschus*, belongs also to the same series.

The balmy and gummy series comprises balsam of Peru, balsam of Tolu, benzoin, styrax, myrrh, and camphor. With the exception of the last, they are all exudations from various trees; balsam of Peru being obtained from the *Myroxylon Peruiferum*, balsam of Tolu from the *Toluifer balsamum*, benzoin (or gum-benjamin) from the *Styrax benzoin*, and myrrh from the *Balsamodendron myrrha*. The four first-named possess a fragrance somewhat similar to vanilla, but less delicate. Myrrh was the most esteemed perfume in ancient times, but tastes must have changed since, for it is now but little in request, and then only for dentifrices. Camphor, which is more used in medicine than perfumery, is obtained by boiling the wood of the *Laurus camphora*, a tree found principally in China and Japan, and in which the gum exists ready formed.

The fruity series includes bitter almonds, Tonquin beans and vanilla. The essential oil of bitter almonds is obtained by distilling the dry cake of the fruit after the fat oil has been pressed out. It contains from eight to ten per cent. of prussic acid, which can be removed by re-distilling it over potash.

Tonquin beans are the fruit of the *Dipterix odorata*, a tree which grows in the West Indies and South America.

Vanilla is the bean of a beautiful creeper (*Vanilla planifolia*) which is a native of Mexico, but has lately been introduced into the French island of Réunion, where it thrives admirably. This colony now yields annually more than 12,000 lbs. of the costly perfume, and among the many beautiful specimens shown at the last Exhibition, nine were deemed worthy of medals or of honourable mention. A sort of bastard vanilla, called vanilloes, is obtained from the *Vanilla Pompona*, which is found in the West Indies and Guiana.

Vanilla Plant.

The artificial series comprises all the various flavours produced by chemical combinations. Of these the most extensively used in perfumery is the nitro-benzine, usually called mirbane, or artificial essence of almonds. This is obtained by treating rectified naphtha with nitric acid and sulphuric acid, or sometimes with nitric acid alone. The naphtha is poured slowly through a tube into the acids, decomposition follows, and the essence is found floating on the surface. Artificial essences of lemon and cinnamon have also been produced, but have not been brought to sufficient perfection to be available for practical use. Besides these, artificial essences imitating fruit flavours are manufactured, but principally for making confectionery. The pear essence is an amylic ether; the apple essence, a valerianic ether, containing amyl; and the pine-apple essence, a butyric ether. The whole of these require to be diluted with five or six times their weight of alcohol, to develop their flavour.

This closes the list of materials used hitherto by perfumers; but there are many other fragrant treasures dispersed all over the globe, which, from want of communication, or the difficulty of extraction, have not yet found their way to our laboratories, but may do so at some future time.

The various floral essences distilled in the East Indies I have noticed in a former chapter. The imperfect way in which they are made, and their very high price, preclude us from making any use of them, but these two obstacles may one day be removed.

In Australia there are many trees with fragrant leaves, and principally the Tasmanian peppermint (*Eucalyptus amygdalina*), the peppermint-tree (*Eucalyptus odorata*) the blue gum-tree (*Eucalyptus globulus*), &c. Essential oils distilled from these leaves were shown at the last Exhibition; and although described in the catalogue as only fit for painting purposes, I expressed an opinion that they might be rendered available for perfumery. An experiment which I made with the oil of *Eucalyptus amygdalina* (possessing a strange flavour of nutmegs combined with peppermint) confirmed me in that idea; and I am pleased to find that colonists have turned their attention to the subject, and are now sending these oils to our markets. The wattle flower is also very abundant in those parts, and as it closely resembles the cassie in fragrance, it might be turned to good account. I received not long since from Tasmania a specimen of pomade made from the flowers of the silver-wattle (*Acacia dealbata*), but it was very inferior, owing to the want of experience in the operator. New South Wales and Queensland produce myall-wood (*Acacia pendula*), which has an intense and delightful smell of violets, a very scarce odour in nature.

Among other novel odorous products shown at the Exhibition, I may mention *Alyxia aromatica*, a fragrant bark from Cochin-China; another bark from New Caledonia, called *Ocotea aromatica*; and a highly-scented wood (*Licoria odorata*) from French Guiana, which has a strong flavour of bergamot.

For the convenience of persons curious or interested in this matter, I have subjoined a table, where all the principal materials used for perfumery are classed in alphabetical order. Besides these, as I said before, there is a vast number of aromatic plants and flowers which have not yet been made available for the per-fumer's art. The Flora of Nice alone furnishes above one hundred and fifty different specimens, of which the

sole mission has been hitherto to embalm the air of
its mountains and valleys, but which, one day, may
be brought into use; for

> " Arabia cannot boast
> A fuller gale of joy, than liberal, thence
> Breathes through the sense, and takes the ravish'd soul."

APPENDIX.

PRINCIPAL MATERIALS USED IN PERFUMERY.

NAMES.	WHENCE EXTRACTED.	PLACE OF PRODUCTION.
Almond (bitter)	Amygdalus amara	Northern Africa.
Ambergris . .	Secretion of the Physeter macro-cephalus	Found floating on the sea, or on the coasts of India, China, Japan, Greenland, and other places.
Aniseed . . .	Pimpinella anisum	North of Europe.
Ditto (Star) . .	Illicium anisatum	China and Japan.
Balsam of Peru	Myroxylon peruiferum	Western coast of South America.
Balsam of Tolu	Toluifera balsamum	Ditto.
Benzoin Gum .	Styrax benzoin	Siam, Sumatra, and Singapore.
Bergamot . .	Citrus Bergamia rind	Calabria and Sicily.
Bigarrade . .	Citrus bigaradia rind	Italy.
Camphor . .	Laurus camphora	China and Japan.
Carraway . .	Carum carui	England, Germany, and France.
Cascarilla . .	Croton cascarilla	Bahama Islands.
Cassia	Laurus Cassia	East Indies and China.
Cassie	Acacia farnesiana	South of France, Italy, Algeria, and Tunis.
Cedar	Pinus Cedra and Juniperus Virginiana	Syria, United States, and Honduras.
Cedrat . . .	Citrus cedrata rind	South of France and Italy.
Cinnamon . .	Laurus Cinnamomum bark . . .	Ceylon.
Cinnamon leaf .	Leaves of the same plant . . .	Ditto.
Citronella . .	Andropogon Citratum	Ditto.
Civet	Secretion of the Viverra Civetta .	Indian Archipelago, and Africa.
Cloves	Flower bud of the Caryophyllus aromaticus	Indian Archipelago, and Zanzibar.
Dill	Anethum graveolens	England.
Fennell . . .	Anethum fœniculum	South of France.
Geranium . .	Pelargonium odoratissimum . .	South of France, Italy, Algeria, and Spain.
Ginger grass .	Andropogon nardus	Ceylon.
Iris or Orris .	Root of the Iris florentina . . .	Italy.
Jasmine . . .	Jasminum odoratissimum . . .	South of France, Italy, Tunis, and Algeria.
Jonquil . . .	Narcissus Jonquila	South of France and Italy.
Laurel . . .	Cerasus lauro-cerasus leaves . .	Ditto.
Lavender . .	Lavandula vera	England, South of France, and Italy.
Lemon . . .	Citrus medica rind	Coast of Genoa, Calabria, Sicily, and Spain.

NAMES.	WHENCE EXTRACTED.	PLACE OF PRODUCTION.
Lemon grass .	Andropogon Schœnanthus . . .	Ceylon.
Limette . . .	Citrus limetta rind	South of France.
Mace	Expressed from the refuse nutmegs	Indian Archipelago.
Marjoram . .	Origana majorana	South of France.
Mirbane . . .	Nitrobenzine or artificial essential oil of almonds	England and France.
Musk	Secretion of the Moschus moschatus	Thibet, China, and Siberia.
Musk seed . .	Hibiscus abelmoschus	West Indies.
Myrtle . . .	Myrtus communis	South of France.
Myrrh . . .	Balsamodendron Myrrha . . .	East Indies and Arabia.
Narcissus . .	Narcissus odorata	Algeria.
Neroli (bigarrade) . .	Citrus Bigaradia flowers . . .	South of France, Italy, and Algeria.
Neroli (Portugal) . .	Citrus aurantium flowers . . .	Ditto.
Nutmeg . . .	Myristica moschata	Indian Archipelago.
Orange or Portugal . .	Citrus aurantium rind	Calabria and Sicily.
Orange flower .	Citrus Bigaradia flowers	South of France and Italy.
Patchouly . .	Pogostemon Patchouli	India and China.
Peppermint . .	Mentha piperita	England and United States.
Petit grain (bigarrade) .	Citrus Bigaradia leaves . . .	South of France and Algeria.
Petit grain (Portugal)	Citrus aurantium leaves	Ditto.
Rose	Rosa centifolia	South of France, Italy and Turkey.
Rosemary . .	Rosmarinus officinalis	South of France.
Rosewood . .	Lignum aspalathum	South America.
Sandalwood . .	Santalum citrinum	India, China, Indian Archipelago, and West Australia.
Sassafras . . .	Laurus sassafrus	United States.
Serpolet . . .	Thymus Serpyllum	South of France.
Spike	Lavandula Spica . . . , . .	Ditto.
Styrax . . .	Liquidambar styraciflua	Turkey.
Thyme . . .	Thymus vulgaris	South of France.
Tonquin . . .	Beans of the Dipterix odorata . .	South America and West Indies.
Tuberose . . .	Polianthes tuberosa	South of France and Italy.
Vanilla . . .	Pod of the Vanilla planifolia . .	Mexico.
Verbena . . .	Aloysia citriodora	Spain.
Violet	Viola odorata	South of France and Italy.
Vitivert . . .	Anatherum muricatum	India.
Wintergreen .	Gaultheria procumbens	United States.

2744114

Made in the USA